From Quarks to Pions

Chiral Symmetry and Confinement

From Quarks to Pions

Chiral Symmetry and Confinement

Michael Creutz

Brookhaven National Laboratory, USA

World Scientific

NEW JERSEY · LONDON · SINGAPORE · BEIJING · SHANGHAI · HONG KONG · TAIPEI · CHENNAI · TOKYO

Published by

World Scientific Publishing Co. Pte. Ltd.

5 Toh Tuck Link, Singapore 596224

USA office: 27 Warren Street, Suite 401-402, Hackensack, NJ 07601

UK office: 57 Shelton Street, Covent Garden, London WC2H 9HE

Library of Congress Cataloging-in-Publication Data
Names: Creutz, Michael, 1944–   author.
Title: From quarks to pions : chiral symmetry and confinement /
    Michael Creutz (Brookhaven National Laboratory, USA).
Description: Singapore ; Hackensack, NJ : World Scientific, [2018] |
    Includes bibliographical references and index.
Identifiers: LCCN 2017056793| ISBN 9789813229235 (hardcover ; alk. paper) |
    ISBN 9813229233 (hardcover ; alk. paper)
Subjects: LCSH: Particles (Nuclear physics)--Chirality. | Quark-gluon interactions. |
    Quark confinement.
Classification: LCC QC793.3.C54 C73 2018 | DDC 539.7/25--dc23
LC record available at https://lccn.loc.gov/2017056793

British Library Cataloguing-in-Publication Data
A catalogue record for this book is available from the British Library.

For any available supplementary material, please visit
http://www.worldscientific.com/worldscibooks/10.1142/10688#t=suppl

Typeset by Stallion Press
Email: enquiries@stallionpress.com

Preface

At a fundamental level, the interaction of quarks with gluon fields lies at the heart of our understanding of the strong nuclear force. However, experimentally we only observe physical hadrons such as protons and pions. This book explores the fascinating physics involved in the path between these contrasting pictures. Along the way symmetries play a crucial role in understanding the parameters of the theory and details of the spectrum of physical particles.

The presentation builds on my review in Ref. [1]. It is meant for readers with a basic understanding of conventional approaches to quantum field theory. The aim is to present a qualitative picture of the importance of non-perturbative effects and how we can understand many features of the strong interactions via symmetry arguments alone.

Many sources have contributed to this interplay of ideas. I am particularly grateful to Ivan Horvath for encouraging me to write the review in Ref. [1] and to Stefano Capitani for a careful preliminary perusal of that review. I am also indebted to my wife, Karen Mack, for her extensive polishing of the text.

Contents

Chapter 1

QCD

The interaction of quarks with gluons is, in some ways, the best understood part of the Standard Model. This quantum field theory is known under the somewhat-whimsical name "Quantum Chromodynamics", or QCD.[1] As will be discussed in this book, we know precisely how to define QCD as the limit of a cutoff theory. On the other hand, the theory has no small parameter; so, perturbation theory, the standard technique of quantum field theory, applies at best only at the highest energies.

This is in striking contrast with the electro-weak sector of the Standard Model. There, the basic electric charge is small and perturbation theory works to incredible accuracy. Nevertheless, the fundamental definition of these interactions remains fuzzy because we know the perturbative series must ultimately diverge.

The treatment of ultraviolet divergences is central to the perturbative approach, requiring a renormalization scheme to obtain physical observables. This is handled rather differently in a non-perturbative scheme such as the lattice. There, the theory is defined as a limiting procedure as the lattice spacing is taken to zero. For QCD, the phenomenon of asymptotic freedom tells us precisely how to take this limit, which occurs as the coupling approaches zero.

Infrared divergences are another annoyance of a perturbative treatment. But QCD, at least with massive quarks, is expected to have a mass gap.

---

[1]If you prefer not to confuse this with the 4000 Angstroms typical of color, you could regard this as an acronym for Quark Confining Dynamics.

Thus, infrared issues should not be relevant. The non-perturbative generation of a mass gap introduces a new scale into the theory. This interplays with the perturbative mass parameters in a non-trivial way, intimately related to issues of gauge field topology.

Perturbation theory often hides crucial qualitative features. Two properties of QCD, confinement and chiral symmetry breaking, stand out as being particularly intractable. Non-perturbative phenomena enter the theory in a fundamental way at both the classical and quantum levels. Over the years, a coherent qualitative picture of the interplay between chiral symmetry, quantum mechanical anomalies, and the lattice has emerged, which forms the theme of this book.

1.1.   Why quarks

Although an isolated quark has not been seen, we have many reasons to believe in the reality of quarks as the basis for this next layer of matter. First, quarks provide a rather elegant explanation of certain regularities in low energy hadronic spectroscopy. It was the successes of the Eight-Fold way [2, 3] which originally motivated the quark model. Two "flavors" of low mass quarks lie at the heart of isospin symmetry in nuclear physics. Adding the somewhat heavier "strange" quark gives the celebrated multiplet structure described by representations of the group $SU(3)$.

Second, the large cross sections observed in deeply inelastic lepton-hadron scattering suggest non-trivial structure within the proton at distance scales of less than 10^{-16} centimeters, whereas the overall proton electromagnetic radius is on the order of 10^{-13} centimeters. Furthermore, the angular dependencies observed in these experiments indicate that any underlying charged constituent carries a half-integer spin [4].

Yet, a further piece of evidence for compositeness lies in the excitations of the low-lying hadrons. Particles differing in angular momentum fall neatly into place along the famous "Regge trajectories" [5]. Families of states group together as orbital excitations of an underlying extended system. The sustained rising of these trajectories with increasing angular momentum points toward strong long-range forces between the constituents.

Finally, the idea of quarks became incontrovertible with the discovery of heavier quark species beyond the first three. The intricate spectroscopy of the charmonium and upsilon families is admirably explained via potential models for non-relativistic bound states. These systems represent what are sometimes thought of as the "hydrogen atoms" of elementary particle physics [6]. The fine details of their structure have since provided a major testing ground for quantitative predictions from lattice techniques.

1.2. Gluons and confinement

Despite its successes, the quark picture raises a variety of puzzles. For the model to work so well, the constituents must not interact so strongly that they lose their identity. The question arises as to whether it is possible to have objects display point-like behavior in a strongly interacting theory. The phenomenon of asymptotic freedom, discussed in more detail later, turns out to be crucial to realizing this picture.

Perhaps the most peculiar aspect of the theory relates to confinement. These basic constituents of matter do not copiously appear as free particles emerging from high energy collisions. This is in marked contrast to the empirical observation in hadronic physics that anything which can be created will be. Only processes prevented by symmetries do not occur. The difficulty in producing quarks has led to the concept of exact confinement. It may be simpler to have a constituent which can never be produced than an approximate imprisonment relying on an unnaturally small suppression factor. This is particularly true in a theory like the strong interaction, which is devoid of any large dimensionless parameters.

But how can one ascribe any reality to an object which cannot be produced? Is this just some sort of mathematical trick? Remarkably, gauge theories potentially possess a simple physical mechanism for giving constituents infinite energy when in isolation. In this picture a quark-antiquark pair experiences an attractive force which remains non-vanishing even for large separations. This linearly-rising long distance potential energy is central to essentially all models of confinement.

For a qualitative description of the mechanism, consider coupling the quarks to a conserved "gluo-electric" flux. In usual electromagnetism the electric field lines spread and give rise to the inverse square law Coulombic field. If one can somehow eliminate massless fields, then a Coulombic spreading will no longer be a solution to the field equations. A Gauss' law constraint states that quarks are the sources of electric fields. If, in removing the massless fields we do not destroy this constraint, the electric lines would be unable to diverge and must form into tubes of conserved flux, schematically illustrated in Fig. 1.1. These tubes begin and end on the quarks and their antiparticles. The flux tube is meant to be a real physical

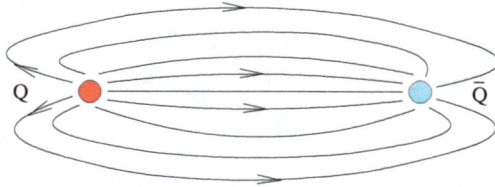

Figure 1.1:   A tube of gluonic flux connects quarks and anti-quarks. The strength of this string is 14 tons.

object carrying a finite energy per unit length. This is the storage medium for the linearly-rising inter-quark potential. In some sense, the reason we cannot have an isolated quark is the same as the reason that we cannot have a piece of string with only one end. In this picture, a baryon would require a string with three ends. It is the group theory of non-Abelian gauge fields that allows this peculiar state of affairs.

Of course a piece of real string can break into two, but then each piece itself has two ends. In the QCD case, a similar phenomenon occurs when there is sufficient energy in the flux tube to create a quark-antiquark pair from the vacuum. This is qualitatively what happens when a rho meson decays into two pions.

One model for this phenomenon is a type II superconductor containing magnetic monopole impurities. Because of the Meissner effect [7], a superconductor does not admit magnetic fields. However, if we force a hypothetical magnetic monopole into the system, its lines of magnetic flux must go somewhere. Here the role of the "gluo-electric" flux is played by the magnetic field, which will bore a tube of normal material through the superconductor until it either ends on an anti-monopole or it leaves the boundary of the system [8]. Such flux tubes have been experimentally observed in real superconductors [9].

Another example of this mechanism occurs in the bag model [10]. Here the gluonic fields are unrestricted in the bag-like interior of a hadron, but are forbidden by *ad hoc* boundary conditions from extending outside. In attempting to extract a single quark from a proton, one would draw out a long skinny bag carrying the gluo-electric flux of the quark back to the remaining constituents.

The above models may be interesting phenomenologically, but they are too arbitrary to be considered as the basis for a fundamental theory. In their search for a more elegant approach, theorists have been drawn to non-Abelian gauge fields [11]. This dynamical system of coupled gluons begins in analogy with electrodynamics, with a set of massless gauge fields interacting with the quarks. Using the freedom of an internal symmetry, the action also includes self-couplings of the gluons. The bare massless fields are all charged with respect to each other. The confinement conjecture is that this input theory of massless charged particles is unstable to a condensation of the vacuum into a state in which only massive excitations can propagate. In such a medium, the gluonic flux around the quarks should form into the flux tubes needed for linear confinement. While this has never been proven analytically, strong evidence from lattice gauge calculations indicates that this is indeed a property of the theory.

The confinement phenomenon makes the theory of the strong interactions qualitatively rather distinct from the theories of the electromagnetic and weak forces. The fundamental fields of the Lagrangian do not manifest themselves in the free particle spectrum. Physical particles are all gauge singlet bound states of the underlying constituents. In particular, an expansion about the free field limit is inherently crippled at the outset. This is perhaps the prime motivation for the lattice approach.

In the quark picture, baryons are bound states of three quarks. Thus, the gauge group should permit singlets to be formed from three objects in the fundamental representation. This motivates the use of $SU(3)$ as the underlying group of the strong interactions. This internal symmetry must not be confused with the broken $SU(3)$ represented in the multiplets of the Eight-Fold way. Ironically, one of the original motivations for quarks has now become an accidental symmetry, arising only because three of the quarks are fairly light. The gauge symmetry of importance to us now is hidden behind the confinement mechanism, which only permits observation of singlet states.

The presentation here assumes, perhaps too naively, that the nuclear interactions can be considered in isolation from the much weaker effects of electromagnetism, weak interactions, and gravitation. This does not preclude the possible application of the techniques to the other interactions. Indeed, unification may be crucial for a consistent theory of the world. At normal laboratory energies, however, it is only for the strong interactions that we are forced to go beyond well-established perturbative methods. Hence, we frame the discussion around quarks and gluons.

Chapter 2

Perturbation theory is not enough

The best evidence for confinement in a non-Abelian gauge theory comes by way of Wilson's [12, 13] formulation on a space-time lattice. This prescription seems a little peculiar at first because the vacuum is not a crystal. Indeed, experimentalists work daily with highly relativistic particles and see no deviations from the continuous symmetries of the Lorentz group. Why, then, have theorists spent so much time describing field theory on the scaffolding of a space-time lattice?

The lattice is a mathematical trick. It provides a cutoff removing the ultraviolet infinities so rampant in quantum field theory. On a lattice, it makes no sense to consider momenta with wavelengths shorter than the lattice spacing. As with any regulator, it must be removed via a renormalization procedure. Physics can only be extracted in the continuum limit, where the lattice spacing is taken to zero. As this limit is taken, the various bare parameters of the theory are adjusted while keeping a few physical quantities fixed at their continuum values.

But infinities and the resulting need for renormalization have been with us since the beginnings of relativistic quantum mechanics. The program for electrodynamics has had immense success without recourse to discrete space. Why reject the time-honored perturbative renormalization procedures in favor of a new cutoff scheme?

Perturbation theory has long been known to have shortcomings in quantum field theory. In a classic paper, Dyson [14] showed that electrodynamics could not be analytic in the coupling around vanishing electric charge. If it were, then one could smoothly continue to a theory where like charges attract rather than repel. This would allow the creation of large separated regions of charge to which additional charges would bind with more energy than their rest masses. This would mean there is no lowest energy state; creating matter-antimatter pairs and separating them into these regions would provide an inexhaustible source of free energy.

2.1. Zero dimensions

The mathematical problems with perturbation theory already appear in the trivial case of zero dimensions. Consider the toy path integral

$$Z(m, g) = \int d\phi \, \exp(-m^2\phi^2 - g\phi^4). \tag{2.1}$$

Formally expanding and naively exchanging the integral with the sum gives

$$Z(m, g) = \sum c_i g^i \tag{2.2}$$

with

$$c_i = \frac{(-1)^i}{i!} \int d\phi \, e^{-m^2\phi^2} \phi^{4i} = \frac{(-1)^i (4i)!}{m^{2i+1} i!}. \tag{2.3}$$

A simple application of Sterling's approximation shows that at large order, these coefficients grow faster than any power. Given any value for g, there will always be an order in the series where the terms grow out of control. Note that by scaling the integrand, we can write

$$Z(m, g) = g^{-1/4} \int d\phi \, \exp(-m^2\phi^2/g^{-1/2} - \phi^4). \tag{2.4}$$

This explicitly exposes a branch cut at the origin, yet another way of seeing the non-analyticity at vanishing coupling.

2.2. Scalar field theories and Landau poles

Thinking non-perturbatively often reveals somewhat surprising results. For example, the ϕ^3 theory of massive scalar bosons coupled with a cubic interaction seems to have a sensible perturbative expansion after renormalization. However, this theory almost certainly does not exist as a quantum system. This is because when the field becomes large, the cubic term in the interaction dominates, and the theory has no minimum energy state. The Euclidean path integral is divergent from the outset since the action is unbounded both above and below.

Perhaps even more surprising, it is widely accepted, although not proven rigorously, that a ϕ^4 theory of bosons interacting with a quartic interaction also does not have a non-trivial continuum limit [15]. The expectation here is that with a cutoff in place, the renormalized coupling will display an upper bound as the bare coupling varies from zero to infinity. If this upper bound then decreases to zero as the cutoff is removed, the renormalized coupling too is driven to zero and we have a free theory.

This issue is sometimes discussed in terms of what is known as the "Landau pole" [16]. In non-asymptotically free theories, such as ϕ^4 and quantum electrodynamics, there is a tendency for the effective coupling to rise with energy. A simple analysis of the lowest order renormalization group equation suggests a possible divergence of the coupling at a finite energy. Not allowing this would force the coupling at smaller energies to zero.[1]

2.3. Solvable theories

The importance of non-perturbative effects is well-understood in a class of two-dimensional models that can be solved via a technique known as "bosonization" [17, 18]. This will be discussed in considerably more detail in Chapter 3. The models that can be solved this way include massless two-dimensional electrodynamics, i.e. the Schwinger model [19], the sine-Gordon model [20], and the Thirring model [21]. These solutions exploit a remarkable mapping between fermionic and bosonic fields in two dimensions. This mapping between commuting and anti-commuting objects is also closely related to the solution to the two-dimensional Ising model [22, 23], discussed in detail in Chapter 4. The Schwinger model in particular has several features in common with QCD. First of all it confines, i.e. the physical "mesons" are bound states of the fundamental fermions. With multiple "flavors" the theory has a natural current algebra [24] and the spectrum in the presence of a small fermion mass has both multiple light "pions" and a heavier eta-prime meson. Finally, the massive theory naturally admits the introduction of a CP-violating parameter, analogous to what we will see for QCD in four dimensions.

---

[1] Of course, as the coupling becomes large, using the lowest order perturbative renormalization flow is completely unjustified. What is more important is whether the full non-perturbative flow has a non-trivial fixed point, as discussed in Chapter 9.

2.4.   QCD

QCD at vanishing coupling constant has free quarks and gluons. This bears no resemblance to the observed physical world of hadrons. As we will see later, renormalization group arguments explicitly demonstrate essential singularities when hadronic properties are regarded as functions of the gauge coupling. This, plus the absence of a small coupling parameter, forces us to go beyond perturbation theory.

There are effects in QCD that are invisible to perturbation theory. The most famous of these is the possibility of having an explicit CP-violating term, usually called theta (Θ). In the classical theory, this involves adding, to the action, a total derivative term that can be rotated away in the perturbative limit. However, as we will discuss extensively later, in the quantum theory there are dramatic physical consequences. Different values of Θ give rise to physically distinct theories that have identical perturbative expansions. One unusual result is that, depending on the parameters of the theory, QCD can even break CP symmetry spontaneously. This is tied to what is known as Dashen's phenomenon [25], first noted even before the days of QCD.

In the mid 1970s, 't Hooft [26] elucidated the underlying connection between this parameter, the chiral anomaly and the topology of gauge fields. Later, Witten [27] used large gauge group ideas to discuss the behavior on Θ in terms of effective Lagrangians. Refs. [28–33] represent a few of the many early studies of the effects of Θ on effective Lagrangians via a mixing between quark and gluonic operators.

Non-perturbative effects also raise subtle questions on the meaning of quark masses. Ordinarily, the mass of a particle is correlated with how it propagates over long distances. This approach fails due to confinement and the fact that a single quark cannot be isolated. With multiple quarks, we will see that there is a complicated non-analytic dependence of the theory on the number of quark species.

The overall picture of the interplay of confinement and chiral symmetry has evolved over many years. The discussion in this book is based on a few reasonably uncontroversial assumptions. First, QCD with N_f light quarks should exist as a field theory and exhibit confinement in the usual way. Then, we assume the validity of the standard picture of chiral symmetry breaking involving a quark condensate $\langle \overline{\psi}\psi \rangle \neq 0$. The conventional chiral perturbation theory based on expanding in masses and momenta around the chiral limit should make sense. We assume the usual result that the anomaly generates a mass for the η' particle, and this mass survives the chiral limit. Throughout, we consider N_f small enough to avoid any potential conformal phase of QCD [34].

2.5.   The lattice as a non-perturbative cutoff

It is the prevalence of non-perturbative phenomena in the strong interactions that forces us to go beyond perturbation theory. This is what has driven us to consider the lattice. The dominant concern is confinement, but issues related to chiral symmetry and quantum mechanical anomalies, to be discussed in later chapters, are also highly non-perturbative. To go beyond the diagrammatic approach, we need a non-perturbative cutoff. Herein lies the main virtue of the lattice, which directly eliminates all wavelengths smaller than the lattice spacing. This occurs before any expansions or approximations are begun.

This situation contrasts sharply with the great successes of quantum electrodynamics, where perturbation theory is central. Most conventional regularization schemes are based on the Feynman expansion; some process is calculated diagrammatically until a divergence is met, and the offending diagram is regulated. Since the basic coupling is so small, only a few terms give good agreement with experiment. While non-perturbative effects are expected, their magnitude is exponentially suppressed in the inverse of the coupling.

On a lattice, a field theory becomes mathematically well-defined and can be studied in various ways. Lattice perturbation theory, although somewhat awkward, should recover all results of other regularization schemes. Discrete space-time, however, allows several alternative approaches. One of these, the strong-coupling expansion, is straightforward to implement. Confinement is automatic in the strong coupling limit because the theory does indeed reduce to one of quarks on the end of strings with finite energy per unit length. While this realization of the flux tube picture provides insight into how confinement can work, unfortunately this limit is not the continuum limit. The latter, as we will see later, involves the weak-coupling limit. To study this, one often turns to numerical simulations, made possible by the lattice reduction of the path integral to a conventional, albeit large, many-dimensional integral.

Chapter 3

Lessons from two dimensions

While not the real world, except to string theorists, field theories in one space dimension and one time direction provide a rich environment for studying field theories beyond perturbation. Many models are exactly solvable due to a fascinating equivalence between bosonic and fermionic theories. This "bosonization" process is possible because there is no spin in one space dimension. The difference between bosons and fermions lies in the transformation properties of the corresponding fields under boosts. With massless particles, chiral symmetry amounts to the fact that no boost can exceed the speed of light and thus a particle moving to, say, the "right", will do so in all frames. In this chapter, we first review the surprising equivalence between a theory of a massless boson and a theory of a massless fermion. Then we consider adding various interactions to solve some models that appear non-trivial in one formulation but are free field theories in the other.

While much of this book involves the use of a lattice cutoff to control infrared and ultraviolet issues, in this chapter we will be a bit more conventional and use a momentum space cutoff. The theories we discuss here are solved by reducing them to free particles. It is primarily in perturbation theory that momentum space is particularly useful, giving diagrams involving loop integrals over internal momenta. However, in a fully non-perturbative context, vastly different scales cannot be fully separated. We will see this particularly in topological features of non-Abelian theories, where both short and long distances become deeply entwined.

Although later we will be concentrating on the Euclidean path integral approach, this chapter works with the Hamiltonian formulation of the relevant field theories. Working in Minkowski space, our sign convention uses

$$a_\mu b_\mu = a_0 b_0 - a_1 b_1$$
$$\epsilon_{0,1} = -\epsilon_{1,0} = 1. \tag{3.1}$$

Chapter 5 will go into how the Hamiltonian and path integral approaches are related via a discretization of time and the transfer matrix formalism. The latter will also appear in Chapter 4, where we solve another two-dimensional lattice theory.

So how is it that one can relate commuting and anti-commuting operators? The trick is to use the fact that with quantum mechanical operators, the conjugate momentum generates "translations" in field space. Consider ordinary quantum mechanics with operators $\hat{x}$ and $\hat{p}$ that satisfy the usual commutation relation

$$[\hat{x}, \hat{p}] = i. \tag{3.2}$$

Then the exponential of $\hat{p}$ translates the coordinate such that

$$e^{i\beta\hat{p}}\hat{x} = (\hat{x} + \beta)e^{i\beta\hat{p}}. \tag{3.3}$$

If we exponentiate $\hat{x}$ as well, we have

$$e^{i\beta\hat{p}}e^{i\beta'\hat{x}} = e^{i\beta\beta'}e^{i\beta'\hat{x}}e^{i\beta\hat{p}}. \tag{3.4}$$

If we select $\beta\beta' = \pi$, then the exponentiated operators anti-commute:

$$[e^{i\beta\hat{p}}, e^{i\beta'\hat{x}}]_+ = 0. \tag{3.5}$$

The remainder of this chapter implements this for one-dimensional field theory.

3.1.   The scalar field

We begin this discussion with the Hilbert space representing the theory of a free scalar particle. This is built up from a vacuum state $|0\rangle$ by applying particle creation operators. There is a pair of creation and annihilation operators a_p and $a_p^\dagger$ for each possible momentum p. As we are working in one space dimension, p is a one-component vector, i.e. a number. These satisfy the commutation relation

$$[a_p, a_{p'}^\dagger] = 4\pi p_0 \delta(p, p'). \tag{3.6}$$

Here $p_0 = \sqrt{p^2 + m^2}$ is the energy of a particle of momentum p. The normalization is only a convention; the factor of p_0 will not be important here, but it does make things transform more nicely under boosts. For now we keep the mass general, although later we will concentrate on the massless case with $p_0 = |p|$. With a normalized vacuum state $\langle 0|0 \rangle = 1$, the one particle states are of the form $|p\rangle = a_p^\dagger |0\rangle$ and are normalized, i.e.

$$\langle p'|p \rangle = 4\pi p_0 \delta(p, p'). \tag{3.7}$$

Many particle states are created by applying multiple creation operators to the vacuum. The Hamiltonian sums the energies of the individual particles:

$$H = \int \frac{dp}{4\pi p_0} \, p_0 a_p^\dagger a_p. \tag{3.8}$$

From these operators we construct our local field and its conjugate momentum

$$\Phi(x) = \int_{-\infty}^{\infty} \frac{dp}{4\pi p_0} \left(e^{-ipx} a_p + e^{ipx} a_p^\dagger \right),$$

$$\Pi(x) = i \int_{-\infty}^{\infty} \frac{dp}{4\pi} \left(e^{-ipx} a_p - e^{ipx} a_p^\dagger \right). \tag{3.9}$$

These satisfy the canonical position space equal-time commutation relations

$$[\Pi(x), \Phi(y)]_{x_0=y_0} = i\delta(x_1 - y_1),$$
$$[\Phi(x), \Phi(y)]_{x_0=y_0} = 0,$$
$$[\Pi(x), \Pi(y)]_{x_0=y_0} = 0. \tag{3.10}$$

In terms of the position space operators, the Hamiltonian is

$$H = \int dx \, \frac{1}{2} : \Pi^2(x) : + \frac{1}{2} : \partial_x \Phi(x)^2 :) + \frac{1}{2} m^2 : \Phi(x)^2 : . \tag{3.11}$$

The colons denote normal ordering with respect to the creation and annihilation operators, i.e. all annihilation operators are placed to the right of all creation operators. This normal ordering ensures a zero energy vacuum.

Commutation relations with the Hamiltonian give the equations of motion for time evolution

$$\frac{d}{dt}\Phi(x) = -i[H, \Phi(x)] = \Pi(x),$$

$$\frac{d}{dt}\Pi(x) = -i[H, \Pi(x)] = \partial_1^2 \Phi(x). \tag{3.12}$$

We can explicitly write the equal time propagator for our scalar field

$$\Delta(x-y) = \langle 0|\Phi(x)\Phi(y)|0\rangle = \int_{-\infty}^{\infty} \frac{dp}{4\pi p_0} e^{-ip(x-y)}. \qquad (3.13)$$

As is well-known [35], a massless field in two dimensions is a rather singular object. In particular, the two-point function $\langle 0|\Phi(x)\Phi(y)|0\rangle$ has an infrared divergence. This can be circumvented by considering correlations between derivatives of the field, which are better behaved. I will shortly introduce infrared and ultraviolet cutoffs, giving well-defined field correlators. Any final conclusions require combinations of the fields having a finite limit, as these cutoff parameters are removed.

3.2.   The fermion field

Now we turn to a free massless fermion. Here I expect to have antiparticles. Use b, c respectively for the particle and antiparticle operators. The anti-commutation relations in Fock space are

$$[b(p), b^{\dagger}(p')]_+ = 4\pi p_0 \delta(p-p'),$$
$$[c(p), c^{\dagger}(p')]_+ = 4\pi p_0 \delta(p-p'),$$
$$[b(p), c^{\dagger}(p')]_+ = 0. \qquad (3.14)$$

For the free massless theory, we want the Hamiltonian to take the form

$$H = \int \frac{dp}{4\pi} (b_p^{\dagger} b_p + c_p^{\dagger} c_p). \qquad (3.15)$$

In position space we construct a two-component fermion field

$$\psi(x) = \int \frac{dp}{4\pi p_0} (e^{-ipx} u(p)b(p) + e^{+ipx} v(p)c^{\dagger}(p), \qquad (3.16)$$

where the two component spinors u and v will be determined shortly. The anti-commutation relations for the field itself become

$$[\psi_i^{\dagger}(x), \psi_j(y)]_+ = \int \frac{dp}{4\pi p_0} e^{ip(x-y)} u_i^{\dagger}(p)u_j(p) + v_i^{\dagger}(-p)v_j(-p). \qquad (3.17)$$

To give the canonical delta function, we want

$$u_i^{\dagger}(p)u_j(p) + v_i^{\dagger}(p)v_j(p) = 2p_0\,\delta_{ij}. \qquad (3.18)$$

This is easily satisfied by taking

$$u = \frac{1}{\sqrt{2}} \begin{pmatrix} \sqrt{p_0 + p} \\ \sqrt{p_0 - p} \end{pmatrix},$$

$$v = \frac{1}{\sqrt{2}} \begin{pmatrix} -\sqrt{p_0 - p} \\ \sqrt{p_0 + p} \end{pmatrix}. \tag{3.19}$$

This leads us to the fermion "propagator"

$$\langle 0|\psi_i^\dagger(x)\psi_j(y)|0\rangle = \int \frac{dp}{4\pi p_0} e^{-ip(x-y)} v^\dagger(p) v(p)$$

$$= \int \frac{dp}{4\pi p_0} e^{-ip(x-y)} (1/2) \begin{pmatrix} p_0 + p & -\sqrt{p_0^2 - p^2} \\ -\sqrt{p_0^2 - p^2} & p_0 - p \end{pmatrix}. \tag{3.20}$$

As the mass vanishes

$$\langle 0|\psi_i^\dagger(x)\psi_j(y)|0\rangle = \int \frac{dp}{4\pi} e^{-ip(x-y)} \begin{pmatrix} \theta(p) & 0 \\ 0 & \theta(-p) \end{pmatrix}$$

$$= \frac{1}{4\pi} \begin{pmatrix} \frac{-i}{x-y-i\epsilon} & 0 \\ 0 & \frac{i}{x-y+i\epsilon} \end{pmatrix}. \tag{3.21}$$

This is what we wish to duplicate with the bosonic operators. Note that at vanishing mass, the upper component represents right movers, and the lower component, the left hand field. This means a natural set of gamma matrices is

$$\gamma_5 = \sigma_3 \qquad \gamma_0 = \sigma_- \qquad \gamma_1 = \sigma_2. \tag{3.22}$$

3.3.   Chiral fields

In one dimension, there is a natural notion of chirality for massless particles. A particle going to the right in one frame does so at the speed of light in all frames. This is true regardless of whether the particle is a boson or a fermion.

Because of this, we can separate the field into right and left moving parts

$$\Phi(x) = \Phi_R(x) + \Phi_L(x), \qquad (3.23)$$

where the right handed piece only involves operators for positive momentum

$$\Phi_R(x) = \int_0^\infty \frac{dp}{4\pi p_0} \left(e^{-ipx} a_p + e^{ipx} a_p^\dagger\right). \qquad (3.24)$$

Correspondingly, the left handed field only involves negative momentum.

Note that the canonical momentum satisfies

$$\Pi(x) = -\partial_x(\Phi_R(x) - \Phi_L(x)). \qquad (3.25)$$

Thus one can alternatively work with $\{\Pi(x), \Phi(x)\}$ or $\{\Phi_R(x), \Phi_L(x)\}$ as a complete set of operators in the Hilbert space. Also note that formally, $\Phi_R(x)$ does not commute with itself at different positions. However, derivatives of the field do, and, as mentioned above, only derivatives of the field are physically sensible. With this proviso, either the left or right fields define a relativistic quantum field theory on its own. Were a mass present, the left and right fields would mix under Lorentz transformations and should not be considered independently.

Under the Hamiltonian of Eq. (3.11), the equations of motion for the chiral fields are particularly simple. The right (left) field only creates right (left) moving waves. In equations, this reads

$$(\partial_t + \partial_x)\Phi_R(x) = 0,$$
$$(\partial_t - \partial_x)\Phi_L(x) = 0. \qquad (3.26)$$

For the time being, concentrate on the right handed field.

As with the full field, correlation functions of these fields are divergent. To get things under better control, introduce an infrared cutoff m and an ultraviolet cutoff ϵ with the definition

$$\Phi_R(x) = \int_m^\infty \frac{dp\, e^{-\epsilon p/2}}{4\pi p} \left(e^{-ipx} a_p + e^{ipx} a_p^\dagger\right). \qquad (3.27)$$

Both cutoffs are to be taken to zero at the end of any calculation of physical relevance.

There is some arbitrariness in both these cutoffs. In particular, another popular infrared cutoff gives the scalar boson a small physical mass via the choice $p_0 = \sqrt{p^2 + m^2}$. All the following could be done either way. A physical mass, however, complicates the separation of chiral parts since Lorentz transformations will mix them. With the choice taken here, the left and right movers remain independent, although Lorentz transformations will change the cutoff.

With the cutoffs in place, the correlation of two of these operators becomes well-defined:

$$\Delta_R(x-y) = \langle 0|\Phi_R(x)\Phi_R(y)|0\rangle = \int_m^\infty \frac{dp\ e^{-\epsilon p}}{4\pi p} e^{-ip(x-y)}$$

$$= \frac{1}{4\pi}\left(C - \log(x-y-i\epsilon) - \log(m) - \frac{i\pi}{2}\right). \qquad (3.28)$$

Here C is the Gompertz constant [36] divided by e and has the value

$$C = \int_1^\infty \frac{dp\ e^{-p}}{p} = 0.21938\ldots. \qquad (3.29)$$

Note that this "propagator" diverges logarithmically as m goes to zero, although its derivatives do not. For example,

$$\langle 0|\partial_x \Phi_R(x)\partial_y \Phi_R(y)|0\rangle = \frac{-1}{4\pi(x-y-i\epsilon)^2} \qquad (3.30)$$

remains a tempered distribution as ϵ goes to zero.

3.4. Exponentiated fields

As mentioned at the beginning of this chapter, the usual construction of fermionic operators in the bosonization process involves exponentiating fields. The use of chiral fields makes this particularly simple. We construct the right (left) fermion fields from the right (left) boson fields alone.

To keep things mathematically precise, consider the normal-ordered operator with the cutoffs in place

$$: e^{i\beta\Phi_R(x)} : = \exp\left(i\beta \int_m^\infty \frac{dp\, e^{-\epsilon p/2}}{4\pi p} e^{ipx} a_p^\dagger\right) \exp\left(i\beta \int_m^\infty \frac{dp\, e^{-\epsilon p/2}}{4\pi p} e^{-ipx} a_p\right).$$
$$(3.31)$$

A straightforward application of the Baker-Campbell-Hausdorff formula, and using the expression in Eq. (3.13), will tell us how to normal order the product of two of these operators

$$: e^{i\beta\Phi_R(x)} : : e^{i\beta'\Phi_R(y)} : = : e^{i\beta\Phi_R(x)+i\beta'\Phi_R(y)} : \ \exp(-\beta\beta'\Delta_R(x-y))$$

$$=: e^{i\beta\Phi_R(x)}e^{i\beta'\Phi_R(y)} : \left(\frac{-ie^C}{m(x-y-i\epsilon)}\right)^{-\beta\beta'/4\pi}. \qquad (3.32)$$

We will always be working with $\beta\beta'$, an integer multiple of 4π; thus, there is no phase ambiguity. This will be the key relation in the following.

In two dimensions, the free massless fermion propagator is proportional to $1/(x - y)$. This is the basis of the bosonization process, which takes $\beta = 2\sqrt{\pi}$ and identifies

$$\psi_R(x) = \frac{e^{C/2}}{\sqrt{2\pi}} \lim_{m \to 0} \sqrt{m} : e^{2i\sqrt{\pi}\Phi_R(x)} :$$

$$\psi_R^\dagger(x) = \frac{e^{C/2}}{\sqrt{2\pi}} \lim_{m \to 0} \sqrt{m} : e^{-2i\sqrt{\pi}\Phi_R(x)} : . \tag{3.33}$$

Using the relation in Eq. (3.32), this gives rise to the desired conventionally normalized fermionic commutation relations for the right handed field.

As they are independent fields, whether the left handed fermion field anti-commutes with the right handed one or not is a convention. To make it anti-commute, multiply $\psi_L(x)$ by minus one raised to the number operator for the right handed particles. This works since the right handed field or its conjugate both change this number by one.[1]

3.5. The mapping

So we can construct fermionic operators entirely from our bosonic fields. The two theories have equivalent Hamiltonians.

$$H_0 = \int \overline{\psi} \gamma_1 \partial_1 \psi \, dx = \int (\Pi^2/2 + (\partial_x \Phi)^2/2) \, dx. \tag{3.34}$$

For the conserved fermion number current, we have

$$j_\mu = \overline{\psi} \gamma_\mu \psi = \frac{1}{\sqrt{\pi}} \epsilon_{\mu\nu} \partial_\nu \Phi. \tag{3.35}$$

Adding an interaction term that is trivial in one formulation can translate into something that looks quite subtle in the other [18]. For example, adding a mass term $m\overline{\psi}\psi$ to the fermion Hamiltonian gives us a theory of free fermions, but this transforms into something proportional to $\sin(\Phi)$ in the boson formulation. This is the famous "sine-Gordon" theory, where a term proportional to $\sin(\Phi)$ added to a free massless boson theory is exactly solvable in terms of free massive fermions [20]. Another solvable model comes from adding a term proportional to $(\partial_x \Phi)^2$ to the free boson Hamiltonian. This is merely a renormalization of the fields and remains

---

[1] This is similar to the convention of whether the electron and proton fields commute or anti-commute.

trivial. However, in terms of the fermion fields, this is proportional to $j_\mu j_\mu$, which looks like a four-fermion coupling. This is the Thirring model [21].

With all this mathematical machinery in place, solving the case of two-dimensional electrodynamics, i.e. the Schwinger model [19], becomes immediately obvious. We take our interaction to be the square of the electric field

$$H_I = \frac{1}{2} \int dx \, E^2 \tag{3.36}$$

where the electric field is obtained by integrating Gauss' law

$$E = g \int_{-\infty}^{x} j_0(x') \, dx', \tag{3.37}$$

with g denoting the electric charge of the fermion. In bosonic language, the current is given in Eq. (3.35); i.e. $j_0 = \frac{1}{\sqrt{\pi}} \partial_1 \Phi$. This makes the integral for the electric field trivial:

$$E = \frac{g\Phi}{\sqrt{\pi}}. \tag{3.38}$$

Thus our interaction Hamiltonian is

$$H_i = \frac{g^2}{2\pi} \Phi^2. \tag{3.39}$$

This is a simple mass term for the phi field.

We have found that two-dimensional electrodynamics with massless fermions is equivalent to a theory of a free massive boson with

$$m = \frac{g}{\sqrt{\pi}}. \tag{3.40}$$

The underlying fermionic degrees of freedom find themselves "confined" into this boson, and this boson has no remaining interactions.

This is a rather remarkable result with analogs for four-dimensional QCD. The theory has acquired a mass gap. Confinement is trivial, occuring because the electric field does not fall off with distance. In four dimensions, the linear potential between quarks is a dynamical question related to whether the gluonic electric fields do spread or are confined into flux tubes, as qualitatively described earlier.

Further study

- Show that the propagator in Eq. (3.13) is indeed not a tempered distribution. Integrate its product by a smooth function, such as a Gaussian, that goes to zero rapidly at infinity. Observe that this integral diverges as the cutoff is removed.

- Solve the two-flavor Schwinger model. Show that the resulting theory corresponds to one massless and one massive boson. Does this theory have an $SU(2)_R \times SU(2)_L$ current algebra [24]?
- Consider adding a small fermion mass term to the Schwinger model. Argue semi-classically that the sign of this mass term is indeed relevant. Similarly, argue that if the mass is sufficiently negative there will be a transition to a phase with a non-trivial expectation of the background electric field. In later chapters, we will see that a similar structure is expected for QCD with one fermion flavor.

Chapter 4

The two-dimensional Ising model

The two-dimensional Ising model [37] is remarkable in that despite having a non-trivial phase transition, the free energy can be calculated exactly. [22, 23] We discuss it here because it provides examples of a variety of tools, many of which are discussed elsewhere in this book. The solution begins with constructing the transfer matrix. This will also be discussed in Chapter 5, and provides the primary pathway between Hamiltonian and Euclidean formulations of quantum systems. The Ising model is converted into a fermion theory, with much paralleling the continuum discussion of Chapter 3. Finally, Fourier transform techniques allow us to diagonalize this operator, directly leading to an explicit expression for the free energy of the model.

To define the model, consider a set of spins $s_{jt} \in \{1, -1\}$ associated with every site of a two-dimensional square lattice. Here, j and t are integers referring to the two dimensions. We will formulate the transfer matrix T to generate translations in the t direction. We can consider the coordinate t as representing "time" and j as "space." For convenience, consider N sites in each direction, ultimately considering the infinite volume limit. We can formally consider boundaries periodic in time and open in space, but because of the large volume limit, these become irrelevant details.

The interaction of the spins is between nearest neighbors. The energy is given by the Hamiltonian

$$E(S) = -J \sum_{j,t} s_{j,t} s_{j+1,t} + s_{j,t} s_{j,t+1}. \tag{4.1}$$

Here, S denotes a general configuration of the spins. The energy is lowered by having neighboring spins being equal. From this, we construct the partition function

$$Z = \sum_{\{s\}} e^{-\beta E(S)}, \tag{4.2}$$

where the sum is taken over all possible spin configurations $\{S\}$. Since β and J always appear as their product, we may select units where $J = 1$ without loss of generality. From the partition function, we can express the free energy per site as

$$F = \frac{1}{N^2} \log(Z). \tag{4.3}$$

To study the phase transition in this system, we are interested in the infinite volume limit, $N \to \infty$.

From the free energy, we can obtain other thermodynamic quantities by differentiation. For example, the internal energy per site is

$$U = \frac{1}{N^2} \langle E \rangle = \frac{\partial}{\partial \beta} F(\beta) \tag{4.4}$$

and the specific heat per site comes from a further derivative

$$C = -\beta^2 \frac{\partial U}{\partial \beta} = -\beta^2 \frac{\partial^2}{\partial^2 \beta} F(\beta). \tag{4.5}$$

In any number of dimensions greater than one, the model exhibits a second order phase transition between a disordered high temperature phase and a magnetized low temperature phase. At this transition, the internal energy is continuous but the specific heat diverges. Here we concentrate on the two-dimensional case, which, as we will see, can be solved exactly for the free energy as a function of the inverse temperature β.

4.1. The transfer matrix

The transfer matrix operates on states representing the spins for all space positions j but a single time t. A generic basis state can be written as $|s_1 \ldots s_N\rangle$ where s_j labels the spins on a given time-slice. We consider these states to be orthonormalized

$$\langle s'_1 \ldots s'_N | s_1 \ldots s_N \rangle = \prod_j \delta_{s'_j s_j}. \tag{4.6}$$

The partition function takes the form $Z = \text{Tr} T^N$. Explicitly, T is defined by its matrix elements between states on successive time-slices

$$\langle s'_1 \ldots s'_N | T | s_1 \ldots s_N \rangle = \exp\left(\beta \sum_j (s_j s_{j+1}/2 + s'_j s'_{j+1}/2 + s_j s'_j) \right).$$

(4.7)

Symmetrization between the successive time-slices conveniently keeps T Hermitian. As the system size $N \to \infty$, we have

$$Z = \sum \lambda_i^N \to \lambda_0^N,$$

(4.8)

where λ_i are the eigenvalues of T with λ_0 being the largest. Our goal is to determine this eigenvalue. The free energy F per site is then given by $F = \log(\lambda_0)/N$. Corrections are exponentially suppressed in N as $e^{-N \log(\lambda_0/\lambda_1)}$, with λ_1 the next largest eigenvalue.

4.2. Domain boundaries as fermions

The trick involves re-expressing the partition function in terms of domain boundaries. A domain of spins with the same value can be regarded as being bounded by the world lines of excited bonds. We will see that these can be expressed in terms of fermion operators that create and destroy excited bonds. Remarkably, the resulting theory is one of free fermions.

Construct an operator a_i that creates an excited bond between sites j and $j + 1$ when that bond is unexcited. This operator should not change the excitation level of any other bond on the slice. We define this operator by its action on the basis states

$$a_j^\dagger | s_1, \ldots s_j, s_{j+1}, \ldots s_N \rangle = \frac{s_j + s_{j+1}}{2} | s_1, \ldots s_j, -s_{j+1}, \ldots - s_N \rangle.$$

(4.9)

Here, all spins from site $j+1$ onward to the right boundary are flipped. The factor $\frac{s_j + s_{j+1}}{2}$ ensures that the bond is not already excited. The boundary conditions are taken to be open, although this is irrelevant to the infinite volume limit. Correspondingly, when the bond is already excited, define a destruction operator to de-excite it:

$$a_j | s_1, \ldots s_j, s_{j+1}, \ldots s_N \rangle = \frac{s_j - s_{j+1}}{2} | s_1, \ldots s_j, -s_{j+1}, \ldots - s_N \rangle.$$

(4.10)

These operators are defined for $1 \leq i < N$.

As defined, these simple operators satisfy fermionic anti-commutation relations

$$[a_j, a_k^\dagger]_+ = \delta_{jk}, \tag{4.11}$$

$$a_j^2 = a_j^{\dagger^2} = 0. \tag{4.12}$$

The number operator $a_j^\dagger a_j$ tells us whether a particular bond is excited or not

$$a_j^\dagger a_j |s_1, \ldots s_N\rangle = \frac{1 - s_j s_{j+1}}{2} |s_1, \ldots s_N\rangle, \tag{4.13}$$

or simply

$$s_j s_{j+1} = 1 - 2a_j^\dagger a_j \tag{4.14}$$

when applied to specific spin states. Since both sides square to unity, this applies to any function of $s_j s_{j+1}$ and we can write one factor from the transfer matrix

$$e^{\beta s_j s_{j+1}/2} = e^{\beta(1/2 - a_j^\dagger a_j)}. \tag{4.15}$$

To proceed, we need an operator that flips a single spin. To find this, note that $a_j^\dagger + a_j$ flips all spins from site $j + 1$ onwards and gives a factor of s_j, while $a_j^\dagger - a_j$ flips the same spins with the factor being s_{j+1} instead. Thus, the combination

$$f_j \equiv (a_{j-1}^\dagger - a_{j-1})(a_j^\dagger + a_j) = a_{j-1}^\dagger a_j + a_j^\dagger a_{j-1} + a_{j-1}^\dagger a_j^\dagger + a_j a_{j-1} \tag{4.16}$$

simply flips s_j with no phase factor. This operator is Hermitian and squares to unity. The flip operators all commute, i.e. $[f_j, f_k] = 0$.

In general, an unexcited bond gives a factor of e^β to the partition function while an excited one gives $e^{-\beta}$. From this, another factor from the transfer matrix can be written in terms of the flip operator

$$e^{\beta s_j' s_j} = e^\beta + f_j e^{-\beta}. \tag{4.17}$$

This allows us to write the transfer matrix fully in terms of the fermionic operators

$$T = (e^{\beta \sum_j (1/2 - a_j^\dagger a_j)}) \left[\prod_j (e^\beta + f_j e^{-\beta}) \right] (e^{\beta \sum_j (1/2 - a_j^\dagger a_j)}). \tag{4.18}$$

The factors in T do not commute. To combine them, it is useful to first exponentiate the flip term into the form

$$e^\beta + f_j e^{-\beta} = A e^{B f_j}, \tag{4.19}$$

where A and B are to be determined. Using $f^2 = 1$ to expand the exponent, we obtain two equations

$$e^\beta = A \cosh(B), \tag{4.20}$$

$$e^{-\beta} = A \sinh(B). \tag{4.21}$$

This is easily solved to give

$$A = \sqrt{2 \sinh(2\beta)}, \tag{4.22}$$

$$B = -\frac{1}{2} \log(\tanh(\beta)). \tag{4.23}$$

Putting it all together, our transfer matrix becomes

$$T = \left(e^{\beta \sum_j (1/2 - a_j^\dagger a_j)} \right) A^N e^{B \sum_j f_j} \left(e^{\beta \sum_j (1/2 - a_j^\dagger a_j)} \right). \tag{4.24}$$

4.3. Diagonalization in momentum space

To proceed, we introduce a discrete Fourier transform and define

$$\tilde{a}_q = \frac{1}{\sqrt{N}} \sum_j e^{2\pi i q j / N} a_j,$$

$$\tilde{a}_q^\dagger = \frac{1}{\sqrt{N}} \sum_j e^{-2\pi i q j / N} a_j^\dagger. \tag{4.25}$$

The inversion is

$$a_j = \frac{1}{\sqrt{N}} \sum_q e^{-2\pi i q j / N} \tilde{a}_q = \frac{1}{\sqrt{N}} \sum_q e^{2\pi i q j / N} \tilde{a}_{-q},$$

$$a_j^\dagger = \frac{1}{\sqrt{N}} \sum_q e^{2\pi i q j / N} \tilde{a}_q^\dagger = \frac{1}{\sqrt{N}} \sum_q e^{-2\pi i q j / N} \tilde{a}_{-q}^\dagger. \tag{4.26}$$

In these equations, it is convenient to consider the momentum range $-N/2 < q \leq N/2$. The transformed variables also satisfy the fermionic anti-commutation relations

$$[\tilde{a}_q, \tilde{a}_{q'}^\dagger]_+ = \delta_{qq'}$$

$$\tilde{a}_q^2 = (\tilde{a}_q^\dagger)^2 = 0. \tag{4.27}$$

After this change of variables, we have for the number operator:

$$\sum_j a_j^\dagger a_j = \sum_q \tilde{a}_q^\dagger \tilde{a}_q, \tag{4.28}$$

and for the spin flip term:

$$\sum_j f_j = \sum_j (a_{j-1}^\dagger - a_{j-1})(a_j^\dagger + a_j) = \sum_q e^{-2\pi i q/n}(\tilde{a}_q^\dagger - \tilde{a}_{-q})(\tilde{a}_{-q}^\dagger + \tilde{a}_q). \tag{4.29}$$

Combining positive and negative q to obtain a sum over positive momenta alone, we have

$$\sum_j f_j = 2\sum_{q>0}(c_q(\tilde{a}_q^\dagger \tilde{a}_q + \tilde{a}_{-q}^\dagger \tilde{a}_{-q} - 1) - i s_q(\tilde{a}_q^\dagger \tilde{a}_{-q}^\dagger - \tilde{a}_{-q}\tilde{a}_q)). \tag{4.30}$$

Here, to keep the notation under control we have defined

$$s_q = \sin(2\pi q/N), \tag{4.31}$$

$$c_q = \cos(2\pi q/N). \tag{4.32}$$

Momentum space has allowed the factorization of the transfer matrix into pieces involving momenta in pairs q and $-q$. As each mode can be either occupied or unoccupied, we have reduced the problem to a set of independent four by four matrices. The largest eigenvalue of the transfer matrix is the product of the largest eigenvalues of each of these matrices.

Actually, these four by four matrices immediately block-diagonalize to pairs of two by two matrices. This follows since all terms in T are quadratic in the fermion operators and thus, can only change the occupancy by 0 or ± 2. When the occupancy is unity, the contribution of momenta q and $-q$ gives two degenerate eigenvalues of $A^2 = 2\sinh(2\beta)$.

The more interesting case involves the mixing of occupancy zero and two. Using a basis with $\begin{pmatrix} 0 \\ 1 \end{pmatrix}$ representing neither q nor $-q$ occupied, and $\begin{pmatrix} 1 \\ 0 \end{pmatrix}$ for both occupied, we can write

$$\tilde{a}_q^\dagger \tilde{a}_q + \tilde{a}_{-q}^\dagger \tilde{a}_{-q} - 1 \rightarrow \sigma_z, \tag{4.33}$$

$$-i(\tilde{a}_q^\dagger \tilde{a}_{-q}^\dagger - \tilde{a}_{-q}\tilde{a}_q) \rightarrow \sigma_y, \tag{4.34}$$

where the sigmas are the usual Pauli matrices. The result is that for each positive q, we need to diagonalize the matrix

$$T_q = e^{-\beta\sigma_z} A^2 \exp\left(2B(s_q\sigma_y + c_q\sigma_z)\right) e^{-\beta\sigma_z}. \tag{4.35}$$

Now we use

$$e^{\vec{a}\cdot\vec{\sigma}} = \cosh(|a|) + \hat{a}\cdot\vec{\sigma}\sinh(|a|) \tag{4.36}$$

to reduce our matrix to

$$T_q = e^{-\beta\sigma_z} A^2 (\cosh(2B) + \sinh(2B)(s_q\sigma_y + c_q\sigma_z)) e^{-\beta\sigma_z}. \tag{4.37}$$

Eliminating the temporaries A and B using Eq. (4.21),

$$A^2\cosh(2B) = A^2(\cosh^2(B) + \sinh^2(B)) = 2\cosh(2\beta) \tag{4.38}$$

$$A^2\sinh(2B) = 2A^2\sinh(B)\cosh(B) = 2, \tag{4.39}$$

we are left with

$$T_q = 2e^{-\beta\sigma_z}(\cosh(2\beta) + (s_q\sigma_y + c_q\sigma_z))e^{-\beta\sigma_z}. \tag{4.40}$$

Explicitly writing out the matrix we need to diagonalize gives

$$T_q = 2\begin{pmatrix} e^{-2\beta}(\cosh(2\beta + c_q) & -is_q \\ is_q & e^{2\beta}(\cosh(2\beta) - c_q) \end{pmatrix}. \tag{4.41}$$

Going back to the Pauli matrix notation, we have

$$T_q = 2\left(\cosh^2(2\beta) - c_q\sinh(2\beta) + \cosh(2\beta)(\sinh(2\beta) - c_q)\sigma_z + s_q\sigma_y\right). \tag{4.42}$$

Now a matrix of form

$$a_0 + \vec{a}\cdot\vec{\sigma} \tag{4.43}$$

has eigenvalues

$$\lambda_{\pm} = a_0 \pm |\vec{a}|. \tag{4.44}$$

Thus our eigenvalues are

$$\lambda_{\pm} = 2(\cosh^2(2\beta) - c_q\sinh(2\beta)) \tag{4.45}$$

$$\pm 2\sqrt{\cosh^2(2\beta)(\sinh(2\beta) - c_q)^2 + s_q^2}.$$

We now obtain the largest eigenvalue of the full transfer matrix T by multiplying the largest eigenvalues of T_q together

$$\lambda_0 = \exp\left(\sum_{q>0} \log\left(2(\cosh^2(2\beta) - c_q \sinh(2\beta))\right) \right. \tag{4.46}$$

$$\left. +2\sqrt{\cosh^2(2\beta)(\sinh(2\beta) - c_q)^2 + s_q^2}\right).$$

For large N, we replace the sum with an integral using $2\pi q/N \longrightarrow p$

$$F = \frac{\log(\lambda_0)}{N} \longrightarrow \int_0^\pi \frac{dp}{2\pi} \log\left(2(\cosh^2(2\beta) - \cos(p)\sinh(2\beta))\right.$$

$$\left. + 2\sqrt{\cosh^2(2\beta)(\sinh(2\beta) - \cos(p))^2 + \sin^2(p)}\right) \tag{4.47}$$

which is the final result for the free energy.

Note that the two eigenvalues in Eq. (4.45) can become equal only when $p = 0$ and

$$\sinh(2\beta_c) = 1. \tag{4.48}$$

Solving for β_c gives

$$\beta_c = \frac{1}{2}\log(1 + \sqrt{2}) = 0.44068679351\ldots. \tag{4.49}$$

This is the critical point for the model. Note that at this point, all four eigenvalues of our four by four matrix are equal to two.

Note that the eigenvalues that meet at the critical point do so continuously. The transition is second order. We leave this brief introduction to the model at this point, although much more can be said [38].

Further study

- Solve the one-dimensional Ising model for the partition function and correlation length.
- Show that the two-dimensional model is self-dual. Begin by writing for each bond

$$e^{\beta ss'} = \sum_{u=\pm 1} (u + 1)\cosh(\beta)/2 + (u - 1)ss' \sinh(\beta)/2. \tag{4.50}$$

Interchange the sums over u and the spins. Show that the sum over spins constrains the u's, leaving behind an Ising model on the dual lattice.

Chapter 5

Path integrals and statistical mechanics

Throughout much of the remainder of this book we will focus on the Euclidean path integral formulation of QCD. This approach to quantum mechanics reveals deep connections with classical statistical mechanics. Here we will explore this relationship for the simple case of a non-relativistic particle in a potential. Starting with a partition function representing a path integral on an imaginary time lattice, we will see how a transfer matrix formalism reduces the problem to the diagonalization of an operator in the usual quantum mechanical Hilbert space of square-integrable functions [39]. In the continuum limit of the time lattice, we obtain the canonical Hamiltonian. Except for our use of imaginary time, this treatment is identical to that in Feynman's early work [40].

5.1. Discretizing time

We begin with the Lagrangian for a free particle of mass m moving in potential $V(x)$

$$L(x, \dot{x}) = K(\dot{x}) + V(x), \tag{5.1}$$

where $K(\dot{x}) = \frac{1}{2}m\dot{x}^2$ and $\dot{x}$ is the time derivative of the coordinate x. Note the unconventional relative positive sign between the two terms in Eq. (5.1). This is because we formulate the path integral directly in imaginary time. This improves mathematical convergence, yet we are left with the usual Hamiltonian for diagonalization.

For a particle traversing a trajectory $x(t)$, we have the action

$$S = \int dt\, L(\dot{x}(t), x(t)).$$ (5.2)

This appears in the path integral

$$Z = \int (dx) e^{-S}.$$ (5.3)

Here the integral is over all possible trajectories $x(t)$. As it stands, Eq. (5.3) is rather poorly defined. To characterize the possible trajectories, we introduce a cutoff in the form of a time lattice. Putting our system into a temporal box of total length $\mathcal{T}$, we divide this interval into $N = \frac{\mathcal{T}}{a}$ discrete time-slices, where a is the time-like lattice spacing. Associated with the ith such slice is a coordinate x_i. This construction is sketched in Figure 5.1. Replacing the time derivative of x with a nearest-neighbor difference, we reduce the action to a sum

$$S = a \sum_i \left[\frac{1}{2} m \left(\frac{x_{i+1} - x_i}{a} \right)^2 + V(x_i) \right].$$ (5.4)

The integral in Eq. (5.3) is now defined as an ordinary integral over all the coordinates

$$Z = \int \left(\prod_i dx_i \right) e^{-S}.$$ (5.5)

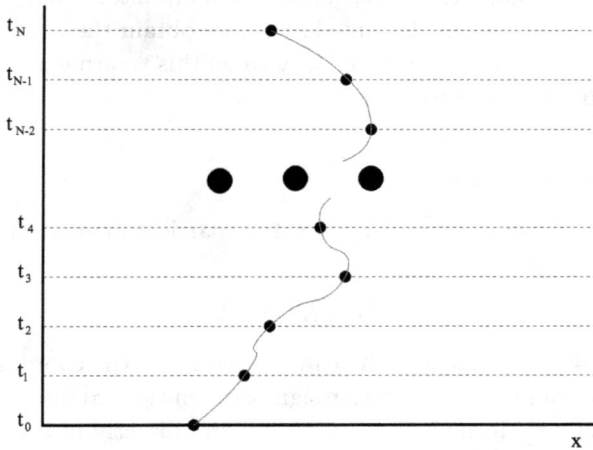

Figure 5.1:   Dividing time into a lattice of N slices of time-step a.

This form for the path integral is precisely in the form of a partition function for a statistical system. We have a one-dimensional polymer of coordinates x_i. The action represents the inverse temperature times the Hamiltonian of the thermal analog. This is a special case of a deep result, a D space-dimensional quantum field theory is equivalent to the classical thermodynamics of a $D + 1$ dimensional system. In this example, we have one degree of freedom and D is zero; for the lattice gauge theory of quarks and gluons, D is three and we work with the classical statistical mechanics of a four-dimensional system.

We now show that the evaluation of this partition function is equivalent to diagonalizing a quantum mechanical Hamiltonian obtained from the action via canonical methods. This is done with the use of the transfer matrix.

5.2. The transfer matrix

The key to the transfer-matrix analysis is to note that the local nature of the action permits us to write the partition function as a matrix product

$$Z = \int \prod_i dx_i \, T_{x_{i+1}, x_i} \tag{5.6}$$

where the transfer-matrix elements are

$$T_{x',x} = \exp\left[-\frac{m}{2a}(x' - x) - \frac{a}{2}(V(x') + V(x))\right]. \tag{5.7}$$

The transfer matrix itself is an operator in the Hilbert space of square-integrable functions with the standard inner product

$$\langle \psi' | \psi \rangle = \int dx \, \psi'^*(x)\psi(x). \tag{5.8}$$

We introduce the non-normalizable basis states $|x\rangle$ such that

$$|\psi\rangle = \int dx \, \psi(x) \, |x\rangle, \tag{5.9}$$

$$\langle x' | x \rangle = \delta(x' - x), \tag{5.10}$$

$$1 = \int dx \, |x\rangle\langle x|. \tag{5.11}$$

Acting on the Hilbert space are the canonically conjugate operators $\hat{p}$ and $\hat{x}$ that satisfy

$$\hat{x}|x\rangle = x|x\rangle$$

$$[\hat{x}, \hat{p}] = i$$

$$e^{-i\hat{p}y}|x\rangle = |x + y\rangle. \tag{5.12}$$

The operator T is defined via its matrix elements

$$\langle x'|T|x\rangle = T_{x',x}, \tag{5.13}$$

where $T_{x',x}$ is given in Eq. (5.7). With periodic boundary conditions on our lattice of N sites, the path integral is compactly expressed as as a trace over the Hilbert space

$$Z = \text{Tr } T^N. \tag{5.14}$$

Expressing T in terms of the basic operators $\hat{p}, \hat{x}$ gives

$$T = \int dy \, e^{-y^2/(2a)} \, e^{-aV(\hat{x})/2} \, e^{-i\hat{p}y} \, e^{-aV(\hat{x})/2}. \tag{5.15}$$

To prove this, check that the right hand side has the appropriate matrix elements. The integral over y is Gaussian and gives

$$T = \left(\frac{2\pi a}{m}\right)^{1/2} e^{-aV(\hat{x})/2} e^{-a\hat{p}^2/(2m)} e^{-aV(\hat{x})/2}. \tag{5.16}$$

The connection with the usual quantum mechanical Hamiltonian appears in the small lattice spacing limit. When a is small, the exponents in the above equation combine to give

$$T = \left(\frac{2\pi a}{m}\right)^{1/2} e^{-aH+O(a^2)} \tag{5.17}$$

with

$$H = \frac{\hat{p}^2}{2m} + V(\hat{x}). \tag{5.18}$$

This is just the canonical Hamiltonian operator following from our starting Lagrangian.

The procedure for going from a path-integral to a Hilbert-space formulation of quantum mechanics consists of three steps. First define the path integral with a discrete time lattice. Then construct the transfer matrix and the Hilbert space on which it operates. Finally, take the logarithm of the transfer matrix and identify the negative of the coefficient of the linear term in the lattice spacing as the Hamiltonian. Physically, the transfer matrix propagates the system from one time-slice to the next. Such time translations are generated by the Hamiltonian.

The eigenvalues of the transfer matrix are related to the energy levels of the quantum system. Denoting the ith eigenvalue of T by λ_i, the path integral or partition function becomes

$$Z = \sum_i \lambda_i^N. \tag{5.19}$$

As the number of time-slices goes to infinity, this expression is dominated by the largest eigenvalue λ_0

$$Z = \lambda_0^N \times [1 + O(\exp[-N \log(\lambda_0/\lambda_1)])]. \tag{5.20}$$

As discussed for the Ising model in Chapter 4, in statistical mechanics the thermodynamic properties of a system follow from this largest eigenvalue. In ordinary quantum mechanics, the corresponding eigenvector is the lowest eigenstate of the Hamiltonian. This is the ground state or, in field theory, the vacuum $|0\rangle$. The connection between imaginary and real time is trivial in this discussion. Whether the generator of time translations is H or iH, we have the same operator to diagonalize.

In statistical mechanics, one is often interested in correlation functions between the fields at different points. This corresponds to a study of the Green's functions of the corresponding field theory. These are obtained upon insertion of polynomials of the fundamental variables into the path integral.

5.3.   Typical paths are non-differentiable

An important feature of the path integral is that a typical path is non-differentiable [41, 42]. Consider the discretization of the time derivative

$$\dot{x} \sim \frac{x_{i+1} - x_i}{a}. \tag{5.21}$$

The kinetic term in the path integral controls how close the fields are on adjacent sites. Since this appears as simple Gaussian factor

$$\exp(-(x_{i+1} - x_i)^2 m/a), \tag{5.22}$$

the average x_{i+1} differs from x_i by something proportional to $\sqrt{a/m}$. This implies

$$\langle \dot{x}^2 \rangle = \langle (x_{i+1} - x_i)^2 \rangle / a^2 = O(1/ma) \qquad (5.23)$$

which diverges as the lattice spacing goes to zero.

One can obtain the physical kinetic energy in other ways, for example through the use of the virial theorem. However, the fact that the typical path is not differentiable means that one should be cautious about generalizing properties of classical fields to typical configurations in a numerical simulation. We will see that such questions naturally arise when considering the topological properties of gauge fields.

Chapter 6

Quark fields and Grassmann integration

Of course, since we are dealing with a theory of quarks, we need additional fields to represent them. There are subtle complications in defining their action on a lattice; we will explore these in some detail in later chapters. Here we go into some of the more mathematical details of what Grassmann integrals mean.

Consider quark fields ψ and $\overline{\psi}$ associated with the sites of the lattice and carrying suppressed spinor, flavor, and color indices. Take a generic action which is a quadratic form in these fields

$$S_f = \overline{\psi}(D + m)\psi. \tag{6.1}$$

We formally separate the kinetic and mass contributions. For the path integral, we are to integrate over ψ and $\overline{\psi}$. It is important that these are independent Grassmann variables. This contrasts strongly with the Hamiltonian case where one usually takes $\overline{\psi} = \psi^\dagger \gamma_0$. Thus ψ and $\overline{\psi}$ on any site anti-commute with each other as well as with ψ and $\overline{\psi}$ on any other site.

Grassmann integration is defined formally as a linear function satisfying a shift symmetry. Consider a single Grassmann variable ψ. Given any function f of ψ, we impose

$$\int d\psi\; f(\psi) = \int d\psi\; f(\psi + \chi) \tag{6.2}$$

where χ is another fixed Grassmann variable. Since the square of any Grassmann variable vanishes, we can expand f in just two terms

$$f(\psi) = \psi a + b. \tag{6.3}$$

Assuming linearity on inserting this into Eq. (6.2) gives

$$\left(\int d\psi \, \psi \right) a + \left(\int d\psi \, 1 \right) b = \left(\int d\psi \, \psi \right) a + \left(\int d\psi \, 1 \right) (a\chi + b). \tag{6.4}$$

This immediately tells us $\int d\psi \, 1$ must vanish. The normalization of $\int d\psi \, \psi$ is still undetermined; the convention is to take this to be unity. Thus the basic Grassmann integral of a single variable is completely determined by

$$\int d\psi \, \psi = 1,$$
$$\int d\psi \, 1 = 0. \tag{6.5}$$

Note that the rule for Grassmann integration seems quite similar to what one would want for differentiation. Indeed, it is natural to define derivatives as anti-commuting objects that satisfy

$$\frac{d}{d\psi} \psi = 1,$$
$$\frac{d}{d\psi} 1 = 0. \tag{6.6}$$

These are exactly the same rules as in Eq. (6.5) for integration. For Grassmann variables, integration and differentiation are the same thing. For the path integral, it is natural to keep the analogy with bosonic fields and refer to integration. On the other hand, for both fermions and bosons we refer to differentiation when using sources in the path integral as a route to correlation functions.

We can make changes of variables in a Grassmann integration in a similar way to ordinary integrals. For example, if we want to change from ψ to $\chi = a\psi$, the above integration rules imply

$$\int d(\psi) f(a\psi) = a \int d(\chi) f(\chi), \tag{6.7}$$

or simply $d(a\psi) = d\chi = \frac{1}{a}d\psi$. We see that the primary difference from ordinary integration is that the Jacobian is inverted. If we consider a multiple integral and take $\chi = M\psi$ with M being a matrix, the transformation generalizes to

$$dx = d(M\psi) = \frac{1}{\det(M)}\, d\psi. \tag{6.8}$$

A particularly important consequence is that we can formally evaluate the Gaussian integrals that appear in the path integral as

$$\int d\psi d\overline{\psi}\, \exp\left(\overline{\psi}(D+m)\psi\right) = \frac{1}{\det(D+m)} = \det\left((D+m)^{-1}\right). \tag{6.9}$$

The normalization is fixed by the earlier conventions. Remember that in the path-integral formulation, ψ and $\overline{\psi}$ represent independent Grassmann fields; in the next section we will discuss the connection between these and the canonical anti-commutation relations for fermion creation and annihilation operators in a quantum mechanical Hilbert space.

In practice, Eq. (6.9) allows one to replace fermionic integrals with ordinary commuting fields ϕ and $\overline{\phi}$ as

$$\int d\psi d\overline{\psi}\, \exp\left(\overline{\psi}(D+m)\psi\right) \propto \int d\phi d\overline{\phi}\, \exp\left(\overline{\phi}(D+m)^{-1}\phi\right). \tag{6.10}$$

This forms the basis for most Monte Carlo algorithms, although the need to invert the large matrix $D+m$ makes such simulations extremely computationally intensive. This approach is, however, still much less demanding than any known way to deal directly with the Grassmann integration in path integrals [43].

6.1. Fermionic transfer matrices

The concept of continuity is lost with Grassmann variables. There is no meaning in saying that fermion fields at nearby sites are near each other in Grassmann space. This is closely tied to the doubling issues that we will discuss later. But it also raises interesting complications in relating Hamiltonian quantum mechanics with the Euclidean formulation involving path integrals. Here we will go into how this connection is made with an extremely simple zero-space-dimensional model.

Anti-commutation is at the heart of fermionic behavior. This is true in both the Hamiltonian operator formalism and the Lagrangian path integral, in rather complementary ways. Starting with a Hamiltonian approach, if an operator $a^\dagger$ creates a fermion in some normalized state on the lattice or the continuum, it satisfies the basic relation

$$[a, a^\dagger]_+ \equiv aa^\dagger + a^\dagger a = 1. \tag{6.11}$$

This contrasts sharply with the fields in a path integral, which all anti-commute:

$$[\chi, \chi^\dagger]_+ = 0. \tag{6.12}$$

The connection between the Hilbert space approach and the path integral appears through the transfer matrix formalism. This is straightforward for bosonic fields [39], but for fermions, certain subtleties arise which are related to the doubling issue [44].

To be more precise, consider a single fermion state created by the operator $a^\dagger$, and an antiparticle state created by another operator $b^\dagger$. For an extremely simple model, consider the Hamiltonian

$$H = m(a^\dagger a + b^\dagger b). \tag{6.13}$$

Here m can be thought of as a "mass" for the particle. What we want is an exact path integral expression for the partition function

$$Z = \mathrm{Tr} e^{-\beta H}. \tag{6.14}$$

Of course, since the Hilbert space generated by a and b has only four states, this is trivial to work out: $Z = 1 + 2e^{-\beta m} + e^{-2\beta m}$. However, we want this in a form that easily generalizes to many variables.

The path integral for fermions uses Grassmann variables. We introduce such a pair, χ and $\chi^\dagger$, which will be connected to the operator pair a and $a^\dagger$, and another pair, ξ and $\xi^\dagger$, for b, $b^\dagger$. All the Grassmann variables anti-commute. Integration over any of them is determined by the simple formulas mentioned earlier

$$\int d\chi \, 1 = 0; \qquad \int d\chi \, \chi = 1. \tag{6.15}$$

For notational simplicity, combine the individual Grassmann variables into spinors

$$\psi = \begin{pmatrix} \chi \\ \xi^\dagger \end{pmatrix}; \quad \psi^\dagger = (\chi^\dagger \quad \xi). \tag{6.16}$$

To make things appear still more familiar, introduce a "Dirac matrix"

$$\gamma_0 = \begin{pmatrix} 1 & 0 \\ 0 & -1 \end{pmatrix} \tag{6.17}$$

and the usual

$$\overline{\psi} = \psi^\dagger \gamma_0. \tag{6.18}$$

Then we have

$$\overline{\psi}\psi = \chi^\dagger\chi + \xi^\dagger\xi, \tag{6.19}$$

where the minus sign from using $\xi^\dagger$ rather than ξ in defining ψ is removed by the γ_0 factor. The temporal projection operators

$$P_\pm = \frac{1}{2}(1 \pm \gamma_0) \tag{6.20}$$

arise when one considers the fields at two different locations

$$\chi_i^\dagger\chi_j + \xi_i^\dagger\xi_j = \overline{\psi}_i P_+ \psi_j + \overline{\psi}_j P_- \psi_i. \tag{6.21}$$

The indices i and j will soon label the ends of a temporal hopping term; this formula is the basic transfer matrix justification for the Wilson projection operator formalism that we return to in later chapters.

6.2. Normal ordering and path integrals

Ignore the antiparticles for a moment and consider some general operator $f(a, a^\dagger)$ in the Hilbert space. How is this related to an integration in Grassmann space? To proceed, we need a convention for ordering the operators in f. We adopt the usual normal ordering definition with the notation $: f(a, a^\dagger) :$ meaning that creation operators are placed to the left of destruction operators, with a minus sign inserted for each exchange. In this case, a rather simple formula gives the trace of the operator as a Grassmann integration

$$\text{Tr} \ : f(a, a^\dagger) : \ = \int d\chi d\chi^\dagger e^{2\chi^\dagger \chi} f(\chi, \chi^\dagger). \tag{6.22}$$

To verify, just check that all elements of the complete set of operators $\{1, a, a^\dagger, a^\dagger a\}$ work. However, this formula is actually much more general; given a set of many Grassmann variables with one pair associated with each of several fermion states, this immediately generalizes to the trace of any normal ordered operator acting in a many-fermion Hilbert space.

What about a product of several normal-ordered operators? This leads to the introduction of multiple sets of Grassmann variables and the general formula

$$\text{Tr} \left(: f_1(a^\dagger, a) : : f_2(a^\dagger, a) : \ldots : f_n(a^\dagger, a) : \right)$$

$$= \int d\chi_1 \, d\chi_1^* \ldots d\chi_n \, d\chi_n^* \, e^{\chi_1^*(\chi_1 + \chi_n)} e^{\chi_2^*(\chi_2 - \chi_1)} \ldots e^{\chi_n^*(\chi_n - \chi_{n-1})}$$

$$\times f_1(\chi_1^*, \chi_1) f_2(\chi_2^*, \chi_2) \ldots f_n(\chi_n^*, \chi_n). \tag{6.23}$$

The positive sign on χ_n in the first exponential factor indicates the natural occurrence of anti-periodic boundary conditions; i.e. we can define $x_0 = -x_n$. With just one factor, this formula reduces to Eq. (6.22). Note how the "time derivative" terms are "one sided;" this is how doubling is eluded.

This exact relationship provides the starting place for converting our partition function into a path integral. The simplicity of our example Hamiltonian allows this to be done exactly at every stage. First, we break "time" into N number of "slices"

$$Z = \text{Tr} \left(e^{-\beta H/N} \right)^N. \tag{6.24}$$

Now we need normal-ordered factors for the above formula. For this we use

$$e^{\alpha a^\dagger a} = 1 + (e^\alpha - 1)a^\dagger a = \; : e^{(e^\alpha - 1)a^\dagger a} : \tag{6.25}$$

which is true for arbitrary α.[1] This is all the machinery we need to write

$$Z = \int (d\psi d\overline{\psi}) e^S, \tag{6.26}$$

where

$$S = \sum_{n=1}^{N} \overline{\psi}_n (e^{-\beta m/N} - 1)\psi_n + \overline{\psi}_n P_+ \psi_{n-1} + \overline{\psi}_{n-1} P_- \psi_n. \tag{6.27}$$

Here, as above, x_0 is defined as $-x_n$. Note how the temporal hopping terms acquire the projection factors $P_\pm$. These take care of the reverse convention of χ versus ξ in our field ψ. The diagonal term gives the factor $-\beta m/N$ appearing in the Hamiltonian form for the partition function.

It is important to realize that if we consider the action as a generalized matrix connecting fermionic variables

$$S = \overline{\psi} M \psi, \tag{6.28}$$

the matrix M is not symmetric. The upper components propagate forward in time, and the lower components backward. Even though our Hamiltonian was Hermitian, the matrix appearing in the corresponding action is

---

[1] The definition of normal ordering gives $: (a^\dagger a)^2 := 0$.

not. With further interactions such as gauge field effects, the intermediate fermion contributions to a general path integral may not be positive, or even real. Of course the final partition function, being the trace of a positive definite operator, is positive. Keeping the symmetry between particles and antiparticles gives a real fermion determinant; this is equivalent to particle-hole symmetry in statistical mechanics. Then, the determinant is naturally positive for an even number of flavors. We will see in later chapters that some rather interesting things can happen with an odd number of flavors.

For our simple Hamiltonian, this discussion has been exact. The discretization of time adds no approximations since we could do the normal ordering by hand. In general, with spatial hopping or more complex interactions, the normal ordering can produce extra terms of order $1/N^{-2}$. In this case, exact results require a limit of a large number of time-slices, but this is a limit we need anyway in order to reach continuum physics.

Chapter 7

Lattice gauge theory

Lattice gauge theory is currently the dominant path to understanding non-perturbative effects in QCD. As formulated by Wilson, the lattice cutoff is quite remarkable in preserving many of the basic ideas of a gauge theory. But just what is a gauge theory anyway? Indeed, there are many ways to think of what is meant by this concept.

At the most simplistic level, a Yang-Mills [11] theory is nothing but an embellishment of electrodynamics with isospin symmetry. For QCD, isospin is generalized to the gauge group $SU(3)$ and the gauge fields are in the octet representation. Since lattice gauge theory is formulated directly in terms of the underlying gauge group elements, this conceptual point of view is inherent in the approach from the start.

At a deeper level, a gauge theory is a theory of phases acquired by a particle as it passes through space time. In electrodynamics, the interaction of a charged particle with the electromagnetic field is elegantly described by the wave function acquiring a phase from the gauge potential. For a particle at rest, this phase adds to its energy an amount proportional to the scalar potential. The use of group elements on lattice links directly gives this connection; the phase associated with some world-line is the product of these elements along the path in question. For the Yang-Mills theory, the concept of phase is generalized to a rotation in the internal symmetry group.

A gauge theory is also a theory with a local symmetry. Gauge transformations involve arbitrary functions of space time. With QCD we have an independent $SU(3)$ symmetry at each point of space time. With the Wilson action formulated in terms of products of group elements around closed loops, this symmetry remains exact even with the cutoff in place.

In perturbative discussions, the local symmetry forces a gauge fixing to remove a formal infinity coming from integrating over all possible gauges. For the lattice formulation, however, the use of a compact representation for the group elements means that the integration over all gauges becomes finite. To study gauge invariant observables, no gauge fixing is required. Of course gauge fixing can still be done, and must be introduced to study more conventional gauge-variant quantities such as gluon or quark propagators. But physical quantities should be gauge invariant; whether gauge fixing is done or not is irrelevant for their calculation.

One aspect of a continuum gauge theory that the lattice does not respect is how a gauge field transforms under Lorentz transformations [45]. In a continuum theory, the basic vector potential can change under a gauge transformation when transforming between frames. For example, the Coulomb gauge treats time in a special way, and a Lorentz transformation can change which direction represents time. The lattice, of course, breaks Lorentz invariance at the outset.

Here we provide only a brief introduction of the lattice approach to a gauge theory. For more details, turn to one of the several excellent books on the subject [46–50]. We postpone to later chapters a discussion of issues related to lattice fermions. These are more naturally understood after further exploration of the peculiarities that must be manifest in any non-perturbative formulation.

7.1.   Link variables

Lattice gauge theory is closely tied to two of the above concepts; it is a theory of phases and it exhibits an exact local symmetry. It is directly formulated in terms of group elements representing the phases acquired by quarks as they hop around the lattice. The basic variables are phases associated with each link of a four-dimensional space-time lattice. In the non-Abelian case, these variables become elements of the gauge group, i.e. $U_{ij} \in SU(3)$ for the strong interactions. Here, i and j denote neighboring sites connected by the link in question. We suppress the group indices to keep the notation under control. The U's are three by three unitary matrices satisfying

$$U_{ij} = U_{ji}^{-1} = (U_{ji})^{\dagger}. \qquad (7.1)$$

The analogy with continuum vector fields A_{μ} is

$$U_{i,i+e_{\mu}} = e^{iag_0 A_{\mu}}. \qquad (7.2)$$

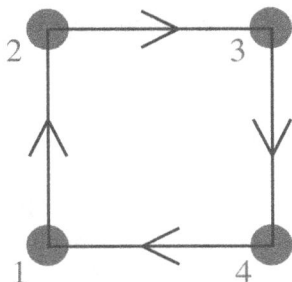

Figure 7.1: Multiplying the group matrices around an elementary square measures the flux through that plaquette. For space-like plaquettes this represents the magnetic field. When one coordinate is time-like, we obtain the electric field.

Here a represents the lattice spacing and g_0 is the bare coupling considered at the scale of the cutoff.

In the continuum, a non-trivial gauge field arises when the curl (in a four-dimensional sense) of the potential is non-zero. This in turn means the phase factor around a small closed loop is not unity. The smallest closed path in the lattice is a "plaquette," or elementary square. Consider the phase corresponding to one such

$$U_P = U_{12}U_{23}U_{34}U_{41} \tag{7.3}$$

where sites 1 through 4 run around the square in question. This is sketched in Fig. 7.1.

In an intuitive sense, this product measures the flux through the plaquette $U_P \sim \exp(ia^2 g_0 F_{\mu,\nu})$, which motivates the use of this quantity to define an action. For this, look at the real part of the trace of U_P

$$\mathrm{Re}\ \mathrm{Tr} U_P = N - a^4 g_0^2\ \mathrm{Tr}\ F_{\mu\nu}F_{\mu\nu} + O(a^6). \tag{7.4}$$

The overall added constant N is physically irrelevant. This leads directly to the Wilson gauge action

$$S(U) = -\sum_P \mathrm{Re}\ \mathrm{Tr} U_P. \tag{7.5}$$

Now we have our gauge variables and an action. To proceed to the quantized theory, we turn to a path integral as an integral over all fields of the exponentiated action. For a Lie group, as will be discussed in the next section, there is a natural measure. Using this measure, the path integral is

$$Z = \int (dU) e^{-\beta S}. \tag{7.6}$$

Here (dU) denotes integration over all link variables. This leads to the usual continuum expression $\frac{1}{2} \int d^4x \operatorname{Tr} F_{\mu\nu} F_{\mu\nu}$ if we choose $\beta = 2N/g_0^2$ for group $SU(N)$ and use the conventionally normalized bare coupling g_0.

Physical correlation functions are obtained from the path integral as expectation values. Given an operator $B(U)$ which depends on the link variables, we have

$$\langle B \rangle = \frac{1}{Z} \int (dU) B(U) e^{-\beta S(U)}. \tag{7.7}$$

Because of the gauge symmetry, this only makes physical sense if B is invariant under gauge transformations.

7.2.   Group integration

The above path integral involves integration over variables which are elements of the gauge group. For this, we use a natural measure with a variety of nice properties. Given any function $f(g)$ of the group elements $g \in G$, the Haar measure is constructed so as to be invariant under "translation" by an arbitrary fixed element g_1 of the group

$$\int dg\, f(g) = \int dg\, f(g_1 g). \tag{7.8}$$

For a compact group, as for the $SU(3)$ relevant to QCD, this is conventionally normalized so that $\int dg\, 1 = 1$. These simple properties are enough for the measure to be uniquely determined.

An explicit representation for this integration measure is almost never needed, but fairly straightforward to write down formally. Suppose a general group element is parameterized by some variables $\alpha_1, \ldots \alpha_n$. Considering here the case $SU(N)$, there are $n = N^2 - 1$ such parameters. Then assume we know some region R in this parameter space that covers the group exactly once. Define the n-dimensional fully anti-symmetric tensor $\epsilon_{i_1, \ldots i_n}$ such that, say, $\epsilon_{1,2,\ldots n} = 1$. Now look at the integral

$$I = A \int_R \{d\alpha\}\, f(g(\vec{\alpha}))\, \epsilon_{i_1, \ldots i_n} \operatorname{Tr}\left((g^{-1}\partial_{i_1} g) \ldots (g^{-1}\partial_{i_n} g)\right). \tag{7.9}$$

This has the required invariance properties of Eq. (7.8). Group properties imply there should be a set of parameters α' depending on α such that $g_1g(\vec{\alpha}) = g(\vec{\alpha}')$. If we change the integration variables from α to α', then the epsilon factor generates exactly the Jacobian needed for this variable change. The normalization factor A is fixed by the above condition $\int dg \ 1 = 1$. Once this is done, we have the invariant measure. The above form for the measure will appear again in Chapter 10 when we discuss topological issues for gauge fields.

Several interesting properties of the Haar measure are easily found. If the group is compact, the left and right measures are equal

$$\int d_R g \ f(g) = \int d_R g \ f(gg_1) = \int d_L g_1 \int d_R g_2 \ f(g_2 g_1) = \int d_L g \ f(g).$$
(7.10)

This also shows the measure is unique, since any left invariant measure could be used. (The normalization can differ for a non-compact group.) A similar argument shows

$$\int dg \ f(g) = \int dg \ f(g^{-1}).$$
(7.11)

For a discrete group, $\int dg$ is simply a sum over the elements. For $U(1) = \{e^{i\theta} | 0 \le \theta < 2\pi\}$ the measure is simply an integral over the circle

$$\int dg \ f(g) = \int_0^{2\pi} \frac{d\theta}{2\pi} f(e^{i\theta}).$$
(7.12)

For $SU(2)$, group elements take the form

$$g = \{a_0 + i\vec{a} \cdot \vec{\sigma} \mid a_0^2 + \vec{a}^2 = 1\}$$
(7.13)

and the measure is

$$\int dg \ f(g) = \frac{1}{\pi^2} \int d^4 a \ f(g)\delta(a^2 - 1).$$
(7.14)

In particular, $SU(2)$ is a 3-sphere.

Some integrals are easily evaluated if we realize that group integration picks out the "singlet" part of a function. Thus

$$\int dg R_{ab}(g) = 0$$
(7.15)

where $R(g)$ is any irreducible matrix representation other than the trivial one, $R = 1$. For the group $SU(3)$, one can write

$$\int dg \ \text{Tr}g \ \text{Tr}g^\dagger = 1$$
(7.16)

$$\int dg \ (\text{Tr}g)^3 = 1$$
(7.17)

from the well-known formulae $3 \otimes \bar{3} = 1 \oplus 8$ and $3 \otimes 3 \otimes 3 = 1 \oplus 8 \oplus 8 \oplus 10$.

A simple integral useful for the strong coupling expansion is

$$\int dg \; g_{ij} \; (g^\dagger)_{kl} = I_{ijkl}. \tag{7.18}$$

The group invariance says we can multiply the indices arbitrarily by a group element on the left or right. There is only one combination of the indices that can survive for $SU(N)$:

$$I_{ijkl} = \delta_{il}\delta_{jk}/N. \tag{7.19}$$

The normalization here is fixed since tracing over jk should give the identity matrix. Another integral that has a fairly simple form is

$$\int dg \; g_{i_1 j_1} \; g_{i_2 j_2} \cdots g_{i_N j_N} = \frac{1}{N!} \epsilon_{i_1 \ldots i_N} \epsilon_{j_1 \ldots j_N}. \tag{7.20}$$

This is useful for studying baryons in the strong coupling regime.

7.3. Gauge invariance

The action of lattice gauge theory has an exact local symmetry. If we associate an arbitrary group element g_i with each site i of the lattice, the action is unchanged if we replace

$$U_{ij} \to g_i^{-1} U_{ij} g_j. \tag{7.21}$$

In constructing the contribution of any plaquette to the action, the gauge transformation factors cancel at the vertices.

One immediate consequence of gauge invariance is that no link can have a vacuum expectation value [51].

$$\langle U_{ij} \rangle = g_i^{-1} \langle U_{ij} \rangle g_j = 0. \tag{7.22}$$

Generalizing this, unless one does some sort of gauge fixing, the correlation between any two separated U matrices is zero. Indeed, many things familiar from perturbation theory often vanish without gauge fixing, including such fundamental objects as quark and gluon propagators!

Another interesting consequence of gauge invariance is that we can forget to integrate over a tree of links in calculating any gauge invariant observable [44]. An axial gauge represents fixing all links pointing in a given direction.[1] Note that this sort of gauge fixing allows the reduction

---

[1] Using a tree with small highly-serrated leaves might be called a "light comb gauge."

of two-dimensional gauge theories to one-dimensional spin models. To see this, pick the tree to be a non-intersecting spiral of links starting at the origin, extending out to the boundary. Links which are transverse to this spiral interact exactly as a one-dimensional system. This also shows that two-dimensional gauge theories are exactly solvable. Construct the transfer matrix along this one-dimensional system. The partition function is the sum of the eigenvalues of this matrix, each raised to the power of the volume of the system.

The trace of any product of link variables around a closed loop is the famous Wilson loop. These quantities are, by construction, gauge invariant and are the natural observables in the lattice theory. The well-known criterion for confinement is whether the expectation of the Wilson loop decreases exponentially with the loop area.

More general gauges can be introduced via an analog of the Fadeev-Popov approach [52]. If $B(U)$ is gauge invariant, then

$$\langle B \rangle = \frac{1}{Z} \int d(U) e^{-S} B(U) = \frac{1}{Z} \int d(U) e^{-S} B(U) f(U) / \phi(U), \quad (7.23)$$

where $f(U)$ is an arbitrary gauge fixing function and

$$\phi(U) = \int (dg) f(g_i^{-1} U_{ij} g_j) \quad (7.24)$$

is the integral of the gauge fixing function f over all gauges. A possible gauge fixing scheme might be to ask that some function h of the links vanishes. In this case we could take $f = \delta(h)$ and then $\phi = \int (dg)\delta(h)$. The integral of a delta function of another function is generically a determinant $\phi = \det(\partial g / \partial h)$. A determinant can generally be written as an integral over a set of auxiliary "ghost" fields. Pursuing this yields the usual Fadeev-Popov picture.

Gauge fixing in the continuum raises subtle issues if one wishes to go beyond perturbation theory. Given some gauge fixing condition $h = 0$ and the corresponding $f = \delta(h)$, it is desirable to have this function vanish only once on any gauge orbit. Otherwise, one should correct for the over-counting due to what are known as "Gribov copies" [53]. This turns out to be non-trivial in most popular perturbative gauges, such as the Coulomb or Landau gauge. For a review, see [54]. One of the great virtues of the lattice approach is that by not fixing the gauge, these issues are sidestepped.

On the lattice, gauge fixing is unnecessary and usually not done if one only cares about measuring gauge invariant quantities such as Wilson loops. But this does have the consequence that the basic lattice fields are far from continuous. The correlation between link variables at different locations vanishes. The locality of the gauge symmetry literally means that there is an independent symmetry at each space-time point. If we consider a quark-antiquark pair located at different positions, they transform under unrelated symmetries. Thus, concepts such as the separation of the potential between quarks into singlet and octet parts are meaningless unless some gauge fixing is imposed.

7.4. Numerical simulation

Monte Carlo simulations of lattice gauge theory have come to dominate the subject. We will introduce some of the basic algorithms in Chapter 8. The idea is to use the analogy with statistical mechanics to generate sets of gauge configurations weighted by the exponentiated action from the path integral. This is accomplished via a Markov chain of small weighted changes to a stored system. Various extrapolations are required to obtain continuum results; the lattice spacing needs to be taken to zero and the lattice size to infinity. Also, such simulations become increasingly difficult as the quark masses become small; thus, extrapolations in the quark mass are generally necessary. It is not the purpose here to cover these techniques in detail; indeed, the several books mentioned at the beginning of this chapter are readily available. In addition, the proceedings of the annual Symposium on Lattice Field Theory are available online for the latest results.

While confinement is natural in the strong coupling limit of the lattice theory [12], we will shortly see that this is not the region of direct physical interest. For this, a continuum limit is necessary. The coupling constant on the lattice represents a bare coupling defined at a length scale given by the lattice spacing. Non-Abelian gauge theories possess the property of asymptotic freedom, which means that in the short distance limit the effective coupling goes to zero. This remarkable phenomenon allows predictions for the observed scaling behavior in deeply inelastic processes. The way quarks expose themselves in high energy collisions was one of the original motivations for a non-Abelian gauge theory of the strong interactions.

In addition to enabling perturbative calculations at high energies, the consequences of asymptotic freedom are crucial for numerical studies via the lattice approach. As the lattice spacing goes to zero, the bare coupling must be taken to zero in a well-determined way. Because of asymptotic freedom, we know precisely how to adjust our simulation parameters to take the continuum limit!

In terms of the statistical analogy, the decreasing coupling takes us away from high temperature and towards the low temperature regime. Along the way a general statistical system might undergo dramatic changes in structure if phase transitions are present. Such qualitative shifts in the physical characteristics of a system can only hamper the task of demonstrating confinement in the non-Abelian theory. Early Monte Carlo studies of lattice gauge theory have provided strong evidence that such troublesome transitions are avoided in the standard four-dimensional $SU(3)$ gauge theory of the nuclear force [55].

Although the ultimate goal of lattice simulations is to provide a quantitative understanding of continuum hadronic physics, many interesting phenomena arise along the way, and they are peculiar to the lattice. Non-trivial phase structure does occur in a variety of models, some of which do not correspond to any continuum field theory. We should remember that when the cutoff is still in place, the lattice formulation is highly non-unique. One can always add additional terms that vanish in the continuum limit. In this way spurious transitions might be alternatively introduced or removed. Going to the continuum limit is required for physical results.

7.5. Order parameters

Formally, lattice gauge theory is like a classical statistical mechanical spin system. The spins U_{ij} are elements of a gauge group G. They are located on the bonds of our lattice. Can this system become "ferromagnetic"? Indeed, as mentioned above, this is impossible since $\langle U \rangle = 0$ follows from the links themselves not being gauge invariant [51].

But we do expect some sort of ordering to occur in the $U(1)$ theory. If this is to describe physical photons, there should be a phase with massless particles. Strong coupling expansions show that for large coupling, this theory has a mass gap [12]. Thus a phase transition is expected, and has been observed in numerical simulations [56]. Exactly how this ordering occurs remains somewhat mysterious; indeed, although people often look for a "mechanism for confinement," it might be interesting to rephrase this question to "How does a theory such as electromagnetism avoid confinement?".

The standard order parameter for gauge theories and confinement involves the Wilson loop mentioned above. This is the trace of the product of link variables multiplied around a closed loop in space-time. If the expectation of such a loop decreases exponentially with the area of the loop, we say the theory obeys an area law and is confining. On the other hand, a decrease only as the perimeter of the loop indicates an unconfined theory. This distinction is inherently non-local; it cannot be measured without involving arbitrarily long distance correlations. The lattice approach is well-known to give the area law in the strong coupling limit of the pure gauge theory. Unfortunately, with dynamical quarks this ceases to be a useful measure of confinement. As a loop becomes large, it will be screened dynamically by quarks "popping" out of the vacuum. Thus in principle, we always have a perimeter law.

Another approach to understanding the confinement phase is to use the mass gap. As long as the quarks themselves are massive, a confining theory should contain no physical massless particles. All mesons, glueballs, and nucleons are expected to gain masses through the dimensional transmutation phenomenon discussed later. As with the area law, the presence of a mass gap is easily demonstrated for the strong coupling limit of the pure glue theory. The deeper question is whether this mass gap persists in the weak coupling limit relevant to the continuum limit.

If the quarks are massless, this definition also becomes a bit tricky. In this case we expect spontaneous breaking of chiral symmetry, also discussed extensively later. This gives rise to pions as massless Goldstone bosons. To distinguish this situation from the unconfined theory, one could consider the number of massless particles in the spectrum by looking at how the "vacuum" energy depends on temperature using the Stefan-Boltzmann law. With N_f flavors we have $N_f^2 - 1$ massless scalar Goldstone bosons. On the other hand, were the gauge group $SU(N)$ not to confine, we would expect $N^2 - 1$ massless vector gauge bosons plus N_f massless quarks, all of which have two degrees of freedom.

Further study

- Consider the two-dimensional Z_2 gauge theory where each link is plus or minus unity. Find an explicit expression for the expectation value of the plaquette.
- Show that the three-dimensional Z_2 gauge theory is dual to the Ising model.

Chapter 8

Monte Carlo simulation

As mentioned earlier, Monte Carlo methods have come to dominate work in lattice gauge theory. These are based on the idea that we need not integrate over all fields, since much information is available already in a few "typical configurations." For bosonic fields these techniques work extremely well, while for fermions the methods remain rather tedious. Over the years, advances in computing power have brought many such calculations for QCD into the realm of possibility. Nevertheless, in situations where the path integral involves complex weightings, the algorithmic issues remain unsolved. In this section we review the basics of the method; this is not meant to be an extensive review, but only a brief introduction.

8.1. Bosonic fields

A generic path integral

$$Z = \int (dU) e^{-S} \tag{8.1}$$

on a finite lattice is a finite dimensional integral. One might try to evaluate it numerically. But it is a many-dimensional integral. With $SU(3)$ on 10^4 lattice we have $4 * 10^4$ links, each parameterized by 8 numbers. Thus, it is a 320,000-dimensional integral. Taking two sample points for each direction, this already gives

$$2^{320,000} = 3.8 \times 10^{96,329} \qquad \text{terms.} \tag{8.2}$$

The age of the universe is only $\sim 10^{27}$ nanoseconds, so adding one term at a time will take a while.

Such big numbers suggest a statistical approach. The goal of a Monte Carlo simulation is to find a few "typical" equilibrium configurations with probability distribution

$$p(C) \sim e^{-\beta S(C)}. \tag{8.3}$$

On these, one can measure observables of choice along with their statistical fluctuations.

The basic procedure is a Markov process

$$C \to C' \to \dots \tag{8.4}$$

generating a chain of configurations that eventually approach the above distribution. In general we take a configuration C to a new one C' with some given probability $P(C \to C')$. As a probability, this satisfies $0 \le P \le 1$ and $\sum_{C'} P(C \to C') = 1$.[1] For a Markov process, P should depend only on the current configuration and have no dependence on the previous history.

The process should bring us closer to "equilibrium." This requires at least two things. First, equilibrium should be stable; *i.e.* equilibrium is an "eigen-distribution" of the Markov chain

$$\sum_{C'} P(C' \to C) e^{-S(C')} = e^{-S(C)}. \tag{8.5}$$

Second, we should have ergodicity; i.e. all possible states must, in principle, be reachable. A remarkable result is that these conditions are sufficient for an algorithm to approach equilibrium, although without any guarantee of efficiency.

Suppose we start with an ensemble of states E characterized by the probability distribution $p(C)$. A distance between ensembles is easily defined

$$D(E, E') \equiv \sum_C |p(C) - p'(C)|. \tag{8.6}$$

This is positive and vanishes only if the ensembles are equivalent. A step of our Markov process takes ensemble E into another E' with

$$p'(C) = \sum_{C'} P(C' \to C) p(C'). \tag{8.7}$$

Now assume that P is chosen so that the equilibrium distribution $p_{eq}(C) = e^{-S(C)}/Z$ is an eigenvector of eigenvalue 1. Compare the new distance from

---

[1] For continuous groups the sum really means integrals.

equilibrium with the old:

$$D(E', E_{eq}) = \sum_C |p'(C) - p_{eq}(C)| = \sum_C \left| \sum_{C'} P(C \to C')(p(C) - p_{eq}(C)) \right|.$$
(8.8)

Now the absolute value of a sum is always less than the sum of the absolute values, so we have

$$D(E', E_{eq}) \leq \sum_C \sum_{C'} P(C \to C')|p(C) - p_{eq}(C)|.$$
(8.9)

Since each C must go somewhere, the sum over C' gives unity and we have

$$D(E', E_{eq}) \leq \sum_C |(p(C) - p_{eq}(C))| = D(E, E_{eq}).$$
(8.10)

Thus the algorithm automatically brings one closer to equilibrium.

How can one be sure that equilibrium is an eigen-ensemble? One common way invokes a principle of detailed balance, a sufficient but not necessary condition. This states that the forward and backward rates between two states are equal when one is in equilibrium, that is,

$$p_{eq}(C)P(C \to C') = p_{eq}(C')P(C' \to C).$$
(8.11)

Summing this over C' immediately gives the fact that the equilibrium distribution is an eigen-ensemble.

The famous Metropolis *et al.* approach [57] is an elegant and simple way to construct an algorithm satisfying detailed balance. This begins with a trial change on the configuration, specified by a trial probability $P_T(C \to C')$. This is required to be constructed in a symmetric way, so that

$$P_T(C \to C') = P_T(C' \to C).$$
(8.12)

This by itself would just tend to randomize the system. To restore the detailed balance, the trial change is conditionally accepted with probability

$$A(C, C') = \min(1, p_{eq}(C')/p_{eq}(C)).$$
(8.13)

In other words, if the Boltzmann weight gets larger, make the change; otherwise, accept it with probability proportional to the ratio of the Boltzmann weights. An explicit expression for the final transition probability is

$$P(C \to C') = P_T(C \to C')A(C, C') + \delta(C, C')$$
$$\times \left(1 - \sum_{C''} P_T(C \to C'')A(C, C'') \right).$$
(8.14)

The delta function accounts for the possibility that the change is rejected.

For lattice gauge theory with its U variables in a group, the trial change can be most easily set up via a table of group elements $T = \{g_1, \ldots g_n\}$. The trial change consists of picking an element randomly from this table and using $U_T = gU$. These can be chosen arbitrarily with two conditions: (1) multiplying them together in various combinations should generate the whole group and (2) for each element in the table, its inverse must also be present, i.e. $g \in T \Rightarrow g^{-1} \in T$. The second condition is essential for having the forward and reverse trial probabilities equal. An interesting feature of this approach is that the measure of the group is not used in any explicit way; it is generated automatically.

Generally the group table should be weighted towards the identity. Otherwise, the acceptance gets small and you never go anywhere. But this weighting should not be too extreme, because then the motion through configuration space becomes slow. Usually the width of the table is adjusted to give an acceptance rate of order 50%. The optimum can be worked out for free field theory — it is a bit less. In general, a big change with a small acceptance can sometimes be better than small changes; this appears to be the case with simulating self-avoiding random walks [58].

The acceptance criterion involves the ratio $\frac{p_{eq}(C')}{p_{eq}(C)}$. An interesting quantity is the expectation of this ratio once in equilibrium. This is

$$\left\langle \frac{p_{eq}(C')}{p_{eq}(C)} \right\rangle = \sum_C p_{eq}(C) \sum_{C'} P_T(C \to C') p_{eq}(C')/p_{eq}(C) = 1$$

(8.15)

since

$$\sum_C P_T(C \to C') = \sum_C P_T(C' \to C) = 1 \qquad (8.16)$$

and $\sum_{C'} p_{eq}(C') = 1$. Monitoring this expectation provides a simple way to follow the approach to equilibrium.

A full Monte Carlo program consists of looping over all the lattice links while considering such tentative changes. To improve performance, there are many tricks that have been developed over the years. For example, in a lattice gauge calculation, the calculation of the "staples" interacting with a given link takes a fair amount of time. This makes it advantageous to apply several Monte Carlo "hits" to the given link before moving on.

8.2. Fermions

The numerical difficulties with fermionic fields stem from their being anti-commuting quantities. Thus it is not immediately straightforward to place them on a computer, which is designed to manipulate numbers. Indeed, the Boltzmann factor with fermions is formally an operator in Grassmann space, and cannot be directly interpreted as a probability. All algorithms in current use eliminate the fermions at the outset by a formal analytic integration. This is possible because most actions in practice are, or can easily be made, quadratic in the fermionic fields. The fermion integrals are then over generalized Gaussians. Unfortunately, the resulting expressions involve the determinant of a large, albeit sparse, matrix. This determinant introduces non-local couplings between the bosonic degrees of freedom, making the path integrals over the remaining fields rather time consuming.

For this brief overview we will be quite generic and assume we are interested in a path integral of the form

$$Z = \int (dA)(d\psi)(d\overline{\psi}) \, \exp\left(-S_G(A) - \overline{\psi}D(A)\psi \right). \tag{8.17}$$

Here the gauge fields are formally denoted A, and fermionic fields, ψ and $\overline{\psi}$. Concentrating on fermionic details in this section, we ignore the technicality that the gauge fields are actually group elements. All details of the fermionic formulation are hidden in the matrix $D(A)$. While we call A a gauge field, the algorithms are general, and have potential applications in other field theories and condensed matter physics.

In the section on Grassmann integration we found the basic formula for a fermionic Gaussian integral

$$\int (d\psi d\overline{\psi}) \, e^{-\overline{\psi}D\psi} = |D| \tag{8.18}$$

where $(d\psi d\overline{\psi}) = d\psi_1 \, d\overline{\psi}_1 \ldots d\psi_n \, d\overline{\psi}_n$. Using this, we can explicitly integrate out the fermions to convert the path integral to

$$Z = \int (dA) \, |D| \, e^{-S_G} = \int (dA) \exp(-S_g + \mathrm{Tr} \, \log(D)). \tag{8.19}$$

This is now an integral over ordinary numbers and therefore, in principle, amenable to Monte Carlo attack.

For now we assume that the fermions have been formulated such that $|D|$ is positive, and thus the integrand can be regarded as proportional to a probability measure. This is true for several of the fermion actions discussed later. However, if $|D|$ is not positive, one can always double the number of fermionic species, replacing D by $D^\dagger D$. We will see in later sections that the case where D is not positive can be rather interesting, but the means by which such situations can be included in numerical simulations is not yet well-understood.

Direct Monte Carlo study of the partition function in this form is still not practical because of the large size of the matrix D. In our compact notation, this is a square matrix of dimension equal to the number of lattice sites times the number of Dirac components times the number of internal symmetry degrees of freedom. Thus, it is typically a hundreds of thousands by hundreds of thousands matrix, precluding any direct attempt to calculate its determinant. It is usually an extremely sparse matrix because most popular actions do not directly couple distant sites. All the Monte Carlo algorithms used in practice for fermions make essential use of this fact.

Some time ago, Weingarten and Petcher [59] presented a simple "exact" algorithm. By introducing "pseudo-fermions" [60, 61] — an auxiliary set of complex scalar fields ϕ — one can rewrite the path integral in the form

$$Z = \int (dA)(d\phi^* \, d\phi) \exp(-S_G - \phi^* D^{-1}\phi). \tag{8.20}$$

Thus, a successful fermionic simulation would be possible if one could obtain configurations of fields ϕ and A with probability distribution

$$P(A,\phi) \propto \exp(-S_G - \phi^* D^{-1}\phi). \tag{8.21}$$

To proceed, we again assume that D is a positive matrix, so this distribution is well-defined.

For an even number of species, generating an independent set of ϕ fields is actually quite easy. If we consider a field χ that is randomly selected and Gaussian in nature, i.e. $P(\chi) \sim e^{-\chi^2}$, then the field $\phi = D\chi$ is distributed as desired for two flavors $P(\phi) \sim e^{-(D^{-1}\phi)^2}$. The hard part of the algorithm is the updating of the A fields, which requires knowledge of how $\phi^* D^{-1}\phi$ changes under trial changes in A.

8.3. The conjugate-gradient algorithm

While D^{-1} is the inverse of an enormous matrix, one really only needs $\phi^* D^{-1}\phi$, which is just one matrix element of this inverse. Furthermore, with a local fermionic action the matrix D is extremely sparse, the non-vanishing matrix elements only connecting nearby sites. In this case there exist quite efficient iterative schemes for finding the inverse of a large sparse matrix applied to a single vector. Here we describe one particularly simple approach.

The conjugate gradient method to find $\xi = D^{-1}\phi$ works by finding the minimum over ξ of the function $|D\xi - \phi|^2$. The solution is iterative; starting with some ξ_0, a sequence of vectors is obtained by moving to the minimum of this function along successive directions d_i. The clever trick of the algorithm is to choose the d_i to be orthogonal in a sense defined by the matrix D itself; in particular $(Dd_i, Dd_j) = 0$ whenever $i \neq j$. This last condition serves to eliminate useless oscillations in undesirable directions and guarantees convergence to the minimum in a number of steps equal to the dimension of the matrix. There are close connections between the conjugate gradient inversion procedure and the Lanczos algorithm for tri-diagonalizing sparse matrices.

The procedure is a simple recursion. Select some arbitrary initial pair of non-vanishing vectors $g_0 = d_0$. For the inversion problem, convergence will be improved if these are a good guess to $D^{-1}\phi$. Then generate a sequence of further vectors by iterating

$$g_{i+1} = (Dg_i, Dd_i)g_i - (g_i, g_i)D^\dagger Dd_i$$

$$d_{i+1} = (Dd_i, Dd_i)g_{i+1} - (Dd, Dg_{i+1})d_i. \tag{8.22}$$

This construction assures that g_i is orthogonal to g_{i+1} and $(Dd_i, Dd_{i+1}) = 0$. It should also be clear that the three sets of vectors $\{d_0, \ldots d_k\}$, $\{g_0, \ldots g_k\}$, and $\{d_0, \ldots (D^\dagger D)^k d_0\}$ all span the same space.

The remarkable core of the algorithm, easily proved by induction, is that the set of g_i are all mutually orthogonal, as are Dd_i. For an N-dimensional matrix, there can be no more than N independent orthogonal vectors. Thus, ignoring round-off errors, the recursion in Eq. (15) must terminate in N or fewer steps with the vectors g and d vanishing from then on. Furthermore, as the above sets of vectors all span the same space, the matrix $D^\dagger D$ is in fact tri-diagonal in a basis defined by the g_i, with (Dg_i, Dg_j) vanishing unless $i = j \pm 1$.

To solve $\phi = D\xi$ for ξ, simply expand in the d_i

$$\xi = \sum_i \alpha_i d_i. \tag{8.23}$$

The coefficients are immediately found from the orthogonality conditions

$$\alpha_i = (Dd_i, \phi)/(Dd_i, Dd_i). \tag{8.24}$$

Note that if we start with the solution $d_0 = D^{-1}\phi$, then we have $\alpha_i = \delta_{i0}$.

This discussion applies for a general matrix D. If D is Hermitian, then one can work with better conditioned matrices by replacing the orthogonality condition for the d_i with (d_i, Dd_j) vanishing for $i \neq j$.

In practice, at least when the correlation length is not too large, this procedure adequately converges in a number of iterations which does not grow severely with the lattice size. As each step involves vector sums with length proportional to the lattice volume, each conjugate gradient step takes a time which grows with the volume of the system. Thus, the overall algorithm including the sweep over lattice variables is expected to require computer time which grows as the square of the volume of the lattice. Such a severe growth has precluded use of this algorithm on any but the smallest lattices. Nevertheless, it does show the existence of an exact algorithm with considerably less computational complexity than would be required for a repeated direct evaluation of the determinant of the fermionic matrix.

Here and below when we discuss volume dependencies, we ignore additional factors from critical slowing down when the correlation length is also allowed to grow with the lattice size. The assumption is that such factors are common for the local algorithms treated here. In addition, such slowing also occurs in bosonic simulations, and we are primarily concerned here with the extra problems presented by the fermions.

8.4.   Hybrid Monte Carlo

One could imagine making trial changes of all lattice variables simultaneously, and then accepting or rejecting the entire new configuration using the exact action. The problem with this approach is that a global random change in the gauge fields will generally increase the action by an amount proportional to the lattice volume, thus the final acceptance rate will fall exponentially with the volume. The acceptance rate could in principle be increased by decreasing the step size of the trial changes, but then the step size would have to decrease with the volume. Exploration of a reasonable region of phase space would thus require a number of steps increasing with the lattice volume. The net result is an exact algorithm which still requires computer time that grows as squared volume.

So far this discussion has assumed that the trial changes are made in a random manner. If, however, one can properly bias these variations, it might be possible to reduce the squared-volume behavior. The "hybrid Monte

Carlo" scheme [62] does this with a global accept/reject step on the entire lattice after a micro-canonical trajectory.

The trick here is to add yet further auxiliary variables in the form of "momentum variables" p conjugate to the gauge fields A. There should be one such variable p_i corresponding to each degree of freedom A_i. Then we look for a coupled distribution

$$P(p, A, \phi) = e^{-H(p,A,\phi)}, \tag{8.25}$$

with

$$H = p^2/2 + V(A) \tag{8.26}$$

and

$$V(A) = -S_g(A) - \phi^* D^{-1} \phi. \tag{8.27}$$

The basic observation is that this is a simple classical Hamiltonian for the conjugate variables A and p, and evolution using Newton's laws will conserve energy. For the gauge fields, one sets up a "trajectory" in a fictitious "Monte Carlo" time variable τ and considers the classical evolution

$$\frac{dA_i}{d\tau} = p_i,$$

$$\frac{dp_i}{d\tau} = F_i(A) = -\frac{\partial V(A)}{\partial A_i}. \tag{8.28}$$

Under such evolution, an equilibrium ensemble will remain in equilibrium.

An approximately energy-conserving algorithm is given by a "leapfrog" discretization of Newton's law. With a micro-canonical time discretization of size δ, this involves two half steps in momentum sandwiching a full step in the coordinate A

$$p_{\frac{1}{2}} = p + \delta\, F(A)/2,$$

$$A' = A + \delta\, p_{\frac{1}{2}},$$

$$p' = p_{\frac{1}{2}} + \delta\, F(A')/2, \tag{8.29}$$

or combined,

$$A' = A + \delta\, p + \delta^2\, F(A)/2,$$

$$p' = p + \delta\, (F(A) + F(A'))/2. \tag{8.30}$$

Even for finite step size δ, this is an area-preserving map of the (A, p) plane onto itself. The scheme iterates this mapping several times before making a final Metropolis accept/reject decision. This iterated map also remains reversible and area-preserving. The most demanding computational part of this process is evaluating the force term. The conjugate gradient algorithm mentioned above can accomplish this.

The important point is that after each step the momentum remains exactly the negative of that which would be required to reverse the entire trajectory and return to the initial variables. If at some point on the trajectory we were to reverse all the momenta, the system would exactly reverse itself and return to the same set of states from whence it came. Thus, a final acceptance with the appropriate probability still makes the overall procedure exact. After each accept/reject step, the momenta p can be refreshed, their values being replaced by new Gaussian random numbers. The pseudo-fermion fields ϕ could also be refreshed at this time. The goal of the procedure is to use the micro-canonical evolution as a way to restrict changes in the action so that the final acceptance will remain high for reasonable step sizes.

This procedure contains several parameters which can be adjusted for optimization. First is N_{mic}, the number of micro-canonical iterations taken before the global accept/reject step and refreshing of the momenta p. Then there is the step size δ, which presumably should be set to give a reasonable acceptance rate. Finally, one can also vary the frequency with which the auxiliary scalar fields ϕ are updated.

The goal of this approach is to speed flow through phase space by replacing a random walk of the A field with a coherent motion in the dynamical direction determined by the conjugate momenta. A simple estimate [63] suggests a net volume dependence proportional to $V^{5/4}$ rather the naive squared volume dependence without these improvements.

As mentioned above, using pseudo-fermions is simplest if the fermion matrix is a square, requiring an even number of species. Users of the hybrid algorithm without the global accept-reject step have argued for adjustment of the number of fermion species by inserting a factor proportional to the number of flavors in front of the pseudo-fermionic term when the gauge fields are updated. This modification is simple to make, but raises some theoretical issues that will be discussed later. In particular, it is crucial that the underlying fermion operator breaks any anomalous symmetries associated with the reduced theory.

Despite the successes of these fermion algorithms, the overall procedure still seems somewhat awkward, particularly when compared with the ease of a pure bosonic simulation. This appears to be tied to the non-local actions resulting from integrating out the fermions. Indeed, had one integrated out a set of bosons coupled quadratically to the gauge field, one would again have a non-local effective action, indicating that this analytic integration was not a good idea. Perhaps we should step back and explore algorithms before integrating out the fermions.

An unsolved problem is to find a practical simulation approach to fermionic systems where the corresponding determinant is not always positive. This situation is of considerable interest because it arises in the study of quark-gluon thermodynamics when a chemical potential is present. All known approaches to this problem are extremely demanding on computer resources. One can move the phase of the determinant into the observables, but then one must divide out the average value of this sign. This is a number which is expected to go to zero exponentially with the lattice volume; thus, such an algorithm will require computer time that grows exponentially with the system size. Another approach is to do an expansion about zero baryon density, which, again to get to large chemical potential, will require rapidly growing resources. New techniques are badly needed to avoid this growth; hopefully this will be a particularly fertile area for future algorithm development.

Chapter 9

Renormalization and the continuum limit

Asymptotic freedom is a signature feature of the theory of the strong interactions. Interactions between quarks decrease at very short distances. From one point of view this allows perturbative calculations in the high energy limit, and this has become an industry in itself. But the concept is also of extreme importance to lattice gauge theory. Indeed, asymptotic freedom tells us precisely how to take the continuum limit. This chapter reviews this crucial connection to the lattice and the renormalization group.

9.1. Coupling constant renormalization

At the level of tree Feynman diagrams, relativistic quantum field theory is well-defined and requires no renormalization. However as soon as loop corrections are encountered, divergences appear and must be removed by a regularization scheme. In general the theory then depends on some cutoff, which is to be removed with a simultaneous adjustment of the bare parameters while keeping physical quantities finite.

For example, consider a lattice cutoff with spacing a. The basic idea is to hold enough physical properties constant to determine how the coupling and quark masses behave as the lattice spacing is reduced. The proton mass m_p is a physical quantity, and on the lattice it will be some, *a priori* unknown, function of the cutoff a, the bare gauge coupling g and the bare quark masses. For the quark-less theory we could use the lightest glueball mass for this purpose.

As the quark masses go to zero the proton mass is expected to remain finite. To simplify the discussion, temporarily ignore the quark masses. Thus, consider the proton mass as a function of the gauge coupling and the cutoff, $m_p(g, a)$. Holding this constant as the cutoff varies determines how g depends on a. This is the basic renormalization group equation

$$a\frac{d}{da}m_p(g(a), a) = 0 = a\frac{\partial}{\partial a}m_p(g, a) + a\left(\frac{dg}{da}\right)\frac{\partial}{\partial g}m_p(g, a). \qquad (9.1)$$

By dimensional analysis, the proton mass should scale as a^{-1} at fixed bare coupling. Thus we know that

$$a\frac{\partial}{\partial a}m_p(g, a) = -m_p(g, a). \qquad (9.2)$$

The "renormalization group function"

$$\beta(g) = a\frac{dg}{da} = \frac{m_p(g, a)}{\frac{\partial}{\partial g}m_p(g, a)} \qquad (9.3)$$

characterizes how the bare coupling is to be varied for the continuum limit. Note that this particular definition is independent of perturbation theory or any gauge fixing.

As renormalization is not needed until quantum loops are encountered, $\beta(g)$ vanishes as g^3 when the coupling goes to zero. Define perturbative coefficients from the asymptotic series

$$\beta(g) = \beta_0 g^3 + \beta_1 g^5 + \cdots. \qquad (9.4)$$

Politzer [64] and Gross and Wilczek [65, 66] first calculated the coefficient β_0 for non-Abelian gauge theories, with the result being

$$\beta_0 = \frac{1}{16\pi^2}(11N/3 - 2N_f/3), \qquad (9.5)$$

where the gauge group is $SU(N)$ and N_f denotes the number of fermion species, or "flavors." As long as $N_f < 11N/2$ this coefficient is positive. Assuming we can reach a region where this first term dominates, decreasing the cutoff corresponds to decreasing the coupling. This is the heart of asymptotic freedom, which tells us that the continuum limit of vanishing lattice spacing requires taking a limit towards vanishing coupling. The two-loop contribution to Eq. (9.4) is also known [67, 68]:

$$\beta_1 = \left(\frac{1}{16\pi^2}\right)^2\left(34N^2/3 - 10NN_f/3 - N_f(N^2 - 1)/N\right). \qquad (9.6)$$

In general, the function $\beta(g)$ depends on the regularization scheme in use. For example, it might depend on what physical property is held fixed as well as the details of how the cutoff is imposed. Remarkably, however, these first two coefficients are universal. Consider two different schemes, each defining a bare coupling as a function of the cutoff (say, $g(a)$ and $g'(a)$). The expansion for one in terms of the other will involve all odd powers of the coupling. In the weak coupling limit, each formulation should reduce to the classical Yang–Mills theory, and thus, to lowest order they should agree:

$$g' = g + cg^3 + O(g^5). \tag{9.7}$$

We can now calculate the new renormalization group function

$$\begin{aligned}
\beta'(g') = a\frac{dg'}{da} &= \frac{\partial g'}{\partial g}\beta(g) \\
&= (1 + 3cg^3)(\beta_0 g^2 + \beta_1 g^3) + O(g^5) \\
&= \beta_0 g'^3 + \beta_1 g'^3 + O(g'^5).
\end{aligned} \tag{9.8}$$

Through order g'^3, the dependence on the parameter c cancels. This, however, does not continue to higher orders, where alternate definitions of the beta function generally differ. Later we will comment further on this non-uniqueness.

The renormalization group function determines how rapidly the coupling decreases with cutoff. Separating variables

$$d(\log(a)) = \frac{dg}{\beta_0 g^3 + \beta_1 g^5 + O(g^7)} \tag{9.9}$$

and integrating allows us to obtain

$$\log(a\Lambda) = -\frac{1}{2\beta_0 g^2} + \frac{\beta_1}{\beta_0^2}\log(g) + O(g^2) \tag{9.10}$$

where Λ is an integration constant. This immediately shows that the lattice spacing decreases exponentially in the inverse coupling

$$a = \frac{1}{\Lambda}e^{-1/2\beta_0 g^2}g^{-\beta_1/\beta_0^2}(1 + O(g^2)). \tag{9.11}$$

Remarkably, although the discussion began with the beta function obtained in perturbation theory, the right hand side of Eq. (9.11) has an essential singularity at vanishing coupling. The renormalization group provides non-perturbative information from a perturbative result.

Dropping the logarithmic corrections, the coupling as a function of the cutoff reduces to

$$g^2 \sim \frac{1}{2\beta_0 \log(1/\Lambda a)}, \tag{9.12}$$

showing the asymptotic freedom result that the bare coupling goes to zero logarithmically with the lattice spacing in the continuum limit.

The integration constant Λ is defined from the bare charge, in a particular cutoff scheme. Its precise numerical value will depend on details, but once the scheme is chosen, it is fixed relative to the scale of the quantity used to define the physical scale. In the above discussion, this was the proton mass. The existence of a scheme dependence can be seen by considering two different bare couplings as related in Eq. (9.7). The relation between the integration constants is

$$\log(\Lambda'/\Lambda) = \frac{c}{2\beta_0}. \tag{9.13}$$

The mass m of a physical particle, perhaps the proton used above, is connected to an inverse correlation length in the statistical analog of the theory. Measuring this correlation length in lattice units, we can consider the dimensionless combination $\xi = 1/am$. For the continuum limit, we want this correlation length to diverge. Multiplying Eq. (9.11) by the mass tells us how this divergence depends on the lattice coupling

$$ma = \xi^{-1} = \frac{m}{\Lambda} e^{-1/2\beta_0 g^2} g^{-\beta_1/\beta_0^2} (1 + O(g^2)). \tag{9.14}$$

Conversely, if we know how a correlation length ξ of the statistical system diverges as the coupling goes to zero, we can read off the particle mass in units of Λ as the coefficient of the behavior in this equation. This exemplifies the close connection between diverging correlation lengths in a statistical system and the continuum limit of the corresponding quantum field theory.

We emphasize again the exponential dependence on the inverse coupling appearing in Eq. (9.14). This is a function that is highly non-analytic at the origin. This demonstrates quite dramatically that QCD cannot be fully described by perturbation theory.

9.2.   A parameter-free theory

This discussion brings us to the remarkable conclusion that, ignoring the quark masses, the strong interactions have no free parameters. The cutoff is absorbed into $g(a)$, which in turn is absorbed into the renormalization group dependence. The only remaining dimensional parameter Λ serves to set the scale for all other masses. When the theory is considered in isolation, one may select units such that Λ is unity. After such a choice, all physical mass ratios are determined. Coleman and Weinberg [69] have given this process, wherein a dimensionless parameter g and a dimensionful one a manage to "eat" each other, the marvelous name "dimensional transmutation."

In the theory including quarks, their masses represent further parameters. Indeed, these are the only parameters in the theory of the strong interactions. In the limit where the quark masses vanish, referred to as the chiral limit, we return to a zero parameter theory. In this approximation to the physical world, the pion mass is expected to vanish and all dimensionless observables should be uniquely determined. This applies not only to mass ratios, such as of the rho mass to the proton, but also to quantities such as the pion-nucleon coupling constant, once regarded as a parameter for a perturbative expansion. As the chiral approximation has been rather successful in the predictions of current algebra, we expect an expansion in the small quark masses to be a fairly accurate description of hadronic physics. Given a qualitative agreement, a fine tuning of the small quark masses should give the pion its mass and complete the theory.

The exciting idea of a parameter-free theory is sadly lacking from most treatments of the other interactions such as electromagnetism or the weak force. There the coupling $\alpha \sim 1/137$ is treated as a parameter. One might optimistically hope that the inclusion of the appropriate non-perturbative ideas into a unified scheme would ultimately render α and the quark and lepton masses calculable.

9.3. Including quark masses

Above, we concentrated on the flow of the bare coupling as one takes the continuum limit. Of course with massive quarks in the theory, the bare quark mass is also renormalized. Here, we extend the above discussion to see how the two bare parameters flow together to zero in a well-defined way.

Including the mass flow, the renormalization group equations become

$$a\frac{dg}{da} = \beta(g) = \beta_0 g^3 + \beta_1 g^5 + \cdots + \text{non-perturbative},$$

$$a\frac{dm}{da} = m\gamma(g) = m(\gamma_0 g^2 + \gamma_1 g^4 + \cdots) + \text{non-perturbative}. \qquad (9.15)$$

Now we have three perturbative coefficients β_0, β_1, γ_0 which are scheme-independent and known [64–68, 70, 71]. For $SU(3)$ we have

$$\beta_0 = \frac{11 - 2N_f/3}{(4\pi)^2} = .0654365977\ldots \qquad (N_f = 1),$$

$$\beta_1 = \frac{102 - 12N_f}{(4\pi)^4} = .0036091343\ldots \qquad (N_f = 1), \qquad (9.16)$$

$$\gamma_0 = \frac{8}{(4\pi)^2} = .0506605918\ldots.$$

For simplicity we work with N_f degenerate quarks, although this is easily generalized to the non-degenerate case. It is important to recognize that the "non-perturbative" parts fall faster than any power of g as $g \to 0$. As we will discuss later, unlike the perturbative pieces, the non-perturbative contributions to γ need not be proportional to the quark mass in general.

As with the pure gauge theory discussed earlier, these equations are easily solved to show

$$a = \frac{1}{\Lambda} e^{-1/2\beta_0 g^2} g^{-\beta_1/\beta_0^2} (1 + O(g^2)),$$

$$m = M g^{\gamma_0/\beta_0} (1 + O(g^2)). \qquad (9.17)$$

The quantities Λ and M are "integration constants" for the renormalization group equations. Rewriting these relations gives the coupling and mass flow in the continuum limit $a \to 0$

$$g^2 \sim \frac{1}{\log(1/\Lambda a)} \to 0 \qquad \text{"asymptotic freedom"},$$

$$m \sim M \left(\frac{1}{\log(1/\Lambda a)} \right)^{\gamma_0/\beta_0} \to 0. \qquad (9.18)$$

Here Λ is usually regarded as the "QCD scale" and M as the "renormalized quark mass." The resulting flow is sketched in Fig. 9.1.

The rate of this flow to the origin is tied to the renormalization group constants, which can be obtained from the inverted equations

$$\Lambda = \lim_{a \to 0} \frac{e^{-1/2\beta_0 g^2} g^{-\beta_1/\beta_0^2}}{a}, \qquad (9.19)$$

$$M = \lim_{a \to 0} m g^{-\gamma_0/\beta_0}. \qquad (9.20)$$

Of course, as discussed for Λ above, the specific numerical values of these parameters depend on the detailed renormalization scheme.

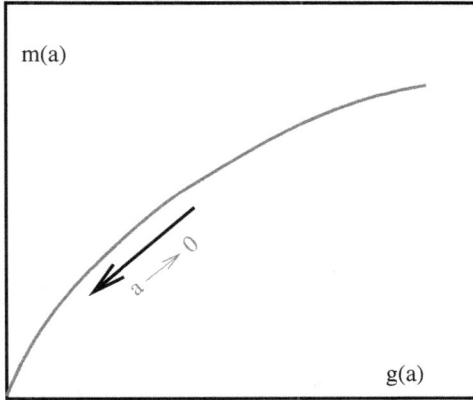

Figure 9.1: In the continuum limit, both the bare coupling and bare mass for QCD flow to zero.

Defining $\beta(g)$ and $\gamma(g)$ is most naturally done by fixing some physical quantities and adjusting the bare parameters as the cutoff is removed. Because of confinement we cannot use the quark mass itself, but we can select several physical particle masses $m_i(g, m, a)$ to hold fixed. This leads to the constraint

$$a\frac{dm_i(g, m, a)}{da} = 0 = \frac{\partial m_i}{\partial g}\beta(g) + \frac{\partial m_i}{\partial m}m\gamma(g) + a\frac{\partial m_i}{\partial a}. \qquad (9.21)$$

For simplicity, continue to work with degenerate quarks of mass m. Then we have two bare parameters (g, m), and we need to fix two quantities.[1] The natural candidates for this are the lightest baryon mass, denoted here as m_p, and the lightest boson mass, m_π. We can rearrange these relations to obtain a somewhat formal but explicit expression for the renormalization group functions

$$\beta(g) = \left(a\frac{\partial m_\pi}{\partial a}\frac{\partial m_p}{\partial m} - a\frac{\partial m_p}{\partial a}\frac{\partial m_\pi}{\partial m}\right) \bigg/ \left(\frac{\partial m_p}{\partial g}\frac{\partial m_\pi}{\partial m} - \frac{\partial m_\pi}{\partial g}\frac{\partial m_p}{\partial m}\right),$$

$$\gamma(g) = \left(a\frac{\partial m_\pi}{\partial a}\frac{\partial m_p}{\partial g} - a\frac{\partial m_p}{\partial a}\frac{\partial m_\pi}{\partial g}\right) \bigg/ \left(\frac{\partial m_p}{\partial m}\frac{\partial m_\pi}{\partial g} - \frac{\partial m_\pi}{\partial m}\frac{\partial m_p}{\partial g}\right).$$

$$(9.22)$$

---

[1] Actually there is a third parameter related to CP conservation. Here we assume CP is a good symmetry and ignore this complication. This issue will be discussed in detail in later chapters.

Note that this particular definition includes all perturbative and non-perturbative effects. In addition, this approach avoids any need for gauge fixing.

Once m_p, m_π, and a renormalization scheme are given, the dependence of the bare parameters on the cutoff is completely fixed. The physical masses are mapped onto the integration constants

$$\Lambda = \Lambda(m_p, m_\pi), \tag{9.23}$$

$$M = M(m_p, m_\pi). \tag{9.24}$$

Formally, these relations can be inverted to express the masses as functions of the integration constants, $m_i = m_i(\Lambda, M)$. Straightforward dimensional analysis tells us that the masses must take the form

$$m_i = \Lambda f_i(M/\Lambda). \tag{9.25}$$

As we will discuss in more detail in later sections, for the multi-flavor theory we expect the pions to be Goldstone bosons with $m_\pi^2 \sim m_q$. This tells us that the above function for the pion should exhibit a square root singularity $f_\pi(x) \sim x^{1/2}$. This relation removes any additive ambiguity in defining the renormalized quark mass M. This conclusion does not persist if the lightest quark becomes non-degenerate.

9.4. Which beta function?

Thus far, our discussion of the renormalization group has been in terms of the bare charge with a cutoff in place. This is the natural procedure in lattice gauge theory; however, there are alternative approaches to the renormalization group that are frequently used in the continuum theory. We now make some comments on the connection between the lattice and the continuum approaches.

An important issue is that there are many different ways to define a renormalized coupling; it should first of all be an observable that remains finite in the continuum limit

$$\lim_{a \to 0} g_r(\mu, a, g(a)) = g_r(\mu). \tag{9.26}$$

Here μ is a dimensionful energy scale introduced to define the renormalized coupling. The subscript r is added to distinguish this coupling from the bare one. For perturbative purposes, one might use a renormalized three-gluon vertex in a particular gauge, with all legs at a given scale of momentum proportional to μ. But many alternatives are possible; for example, one might use as an observable the force between two quarks at separation $1/\mu$.

Secondly, to be properly called a renormalization of the classical coupling, g_r should be normalized such that it reduces to the bare coupling in lowest order perturbation theory for the cutoff theory

$$g_r(\mu, a, g) = g + O(g^3). \tag{9.27}$$

Beyond this, the definition of g_r is totally arbitrary. In particular, given any physical observable H defined at scale μ and satisfying a perturbative expansion

$$H(\mu, a, g) = h_0 + h_1 g^2 + O(g^4), \tag{9.28}$$

we might define a corresponding renormalized coupling

$$g_H^2(\mu) = (H(\mu) - h_0)/h_1. \tag{9.29}$$

As the energy scale goes to infinity, this renormalized charge should go to zero. From this flow of the renormalized charge, we can define a renormalized beta function associated with the observable under consideration,

$$\beta_r(g_r) = -\mu \frac{\partial g_r(\mu)}{\partial \mu}. \tag{9.30}$$

We now draw a remarkable connection between the renormalized $\beta_r(g_r)$ and the function $\beta(g)$ defined earlier for the bare coupling. When the cutoff is still in place, the renormalized coupling is a function of the scale μ of the observable, the cutoff a, and the bare coupling g. Since we are working with dimensionless couplings, g_r can depend directly on μ and a only through their product. This simple application of dimensional analysis implies

$$a \left. \frac{\partial g_r}{\partial a} \right|_g = \mu \left. \frac{\partial g_r}{\partial \mu} \right|_g = -\beta_r. \tag{9.31}$$

In the continuum limit, where we take a to zero and adjust g appropriately, g_r should become a function of the physical scale μ alone. Indeed, we could use $g_r(\mu)$ itself as the physical quantity to hold fixed for the continuum limit. Since it is constant as we take a and g to zero together, we obtain

$$a \frac{\partial g_r}{\partial a} + \frac{\partial g_r}{\partial g} a \frac{\partial g}{\partial a} = 0. \tag{9.32}$$

Using this in an analysis similar to that in Eq. (9.8), we find

$$\beta_r(g_r) = \beta_0 g_r^3 + \beta_1 g_r^5 + O(g_r^7), \qquad (9.33)$$

where β_0 and β_1 are the same coefficients that appear in Eq. (9.4). Both the renormalized and the bare β functions have the same first two coefficients in their perturbative expansions. Indeed, it was through consideration of the renormalized coupling that β_0 and β_1 were first calculated.

It is important to reiterate the considerable arbitrariness in defining both the bare and the renormalized couplings. Far from the continuum, there need not be a simple relationship between different formulations. Once one leaves the perturbative region, even such things as zeros in the β functions are not universal. For an extreme example, the beta function can be forced to consist of only the first two terms.[2] In this case, as long as N_f is small enough that $\beta_1 > 0$, there is explicitly no other zero of the beta function except at $g = 0$.

In contrast, one might think it natural to define the coupling from the force between two quarks. When dynamical quarks are present, screening of the string tension occurs at large distances due to quark pair production. Eventually, the force between the initial quarks will fall exponentially with the pion mass. Then, the beta function becomes formally negative. Thus, this beta function must have a zero in the vicinity of where the screening sets in. We see that even the existence of zeros in the beta function is scheme-dependent. The only exception to this is if a second zero occurs in a region of coupling small enough that perturbation theory can be trusted. This can happen for a sufficient number of flavors [34].

The perturbative expansion of β_r has important experimental consequences. If, as expected, the continuum limit is taken at vanishing bare coupling and the renormalized coupling is small enough that the first terms in Eq. (9.33) dominate, then the renormalized coupling will be driven to zero logarithmically as its defining scale μ goes to infinity. Not only does the bare coupling vanish, but the effective renormalized coupling becomes arbitrarily weak at short distances. This is the physical implication of asymptotic freedom; phenomena involving only short-distance effects may be accurately described with a perturbative expansion. Indeed, asymptotically free gauge theories were first invoked for the strong interactions as an explanation for the apparently free parton behavior manifested in the structure functions associated with deeply inelastic scattering of leptons from hadrons.

---

[2] G. 't Hooft, in [72].

The dependence of the integration constant Λ on the details of the renormalization scheme carries over to the continuum renormalization group as well. Given a particular definition of the renormalized coupling $g_r(\mu)$, its behavior as r goes to zero will involve a scale Λ_r in analogy to the scale in the bare coupling. Hasenfratz and Hasenfratz [73, 74] were the first to perform the necessary one-loop calculations to relate Λ from the Wilson lattice gauge theory with Λ_r defined from the three-gluon vertex in the Feynman gauge and with all legs carrying momentum μ^2. They found

$$\frac{\Lambda_r}{\Lambda} = \begin{pmatrix} 57.5 & SU(2) \\ 83.5 & SU(3) \end{pmatrix} \tag{9.34}$$

for the pure gauge theory. Note that not only is Λ scheme-dependent, but that different definitions can vary by rather large factors. The original calculation of these numbers was rather tedious. They have been verified with calculationally more efficient techniques based on quantum fluctuations around a slowly varying classical background field [75].

9.5. Flows and irrelevant operators

We now briefly discuss another way of looking at the renormalization group as relating theories with different lattice spacings. Given a lattice theory, one could imagine generating an equivalent one with a larger lattice spacing by integrating out a subset of links. While this is conceptually possible, to do it exactly in more than one dimension generates an infinite number of couplings. If we could keep track of such, the procedure would be "exact," but in practice we usually need some truncation. Continuing to integrate out further degrees of freedom, the new couplings flow and might reach some "fixed point" in this infinite coupling-constant space, as sketched in Fig. 9.2. If the fixed point has only one attractive direction, then two different models that flow towards that same fixed point will have the same physics in the large distance limit. This is the concept of universality; exponents are the same for all models with the same attractor.

Some hints on this process come from dimensional analysis, although, in ignoring non-perturbative effects that might occur at strong coupling, the following arguments are not rigorous. In d dimensions, a conventional scalar field has dimensions of $M^{\frac{d-2}{2}}$. Thus, the coupling constant λ in an interaction of form $\int d^d x \, \lambda \phi^n$ has dimensions of $M^{d-n\frac{d-2}{2}}$. On a lattice of spacing a, the natural unit of dimension is the inverse lattice spacing. Thus, without any special tuning, the renormalized coupling at some fixed

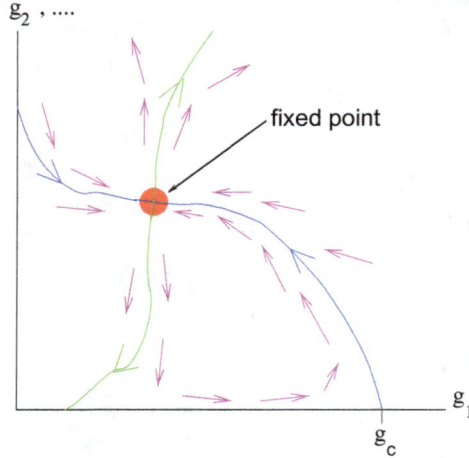

Figure 9.2:   A generic renormalization group flow. In general, this occurs in an infinite dimensional space of coupling constants.

physical scale would naturally run as $\lambda \sim a^{n\frac{d-2}{2}-d}$. As long as the exponent in this expression is positive, i.e.

$$n \geq \frac{2d}{d-2}, \tag{9.35}$$

the coupling will approach zero as a power of a. In this case we expect the coupling to become "irrelevant" in the continuum limit. The fixed point is driven towards zero in the corresponding direction. If d exceeds four, this is the case for all interactions. (We ignore ϕ^3 in 6 dimensions because of stability problems.) This suggests that four dimensions is a critical case, with mean field theory giving the right qualitative critical behavior for all larger dimensions. In four dimensions we have several possible "renormalizable" couplings which are dimensionless, suggesting logarithmic corrections to the simple dimensional arguments. Indeed, four-dimensional non-Abelian gauge theories display exactly such a logarithmic flow; this is asymptotic freedom.

    This simple dimensional argument applied to the mass term suggests it would flow towards infinity in all dimensions. For a conventional phase transition, something must be tuned to a critical point. In statistical mechanics, this is the temperature. In field theory language, we usually remap this onto a tuning of the bare mass term, saying that the transition occurs as bare masses go through zero. For a scalar theory, this tuning for a continuum limit seems unnatural and is one of the unsatisfying features of the Standard Model, driving particle physicists to try to unravel how the Higgs mechanism really works.

In non-Abelian gauge theories with multiple massless fermions, chiral symmetry protects the mass from additive renormalization, avoiding any special tuning. Indeed, as we have discussed, because of dimensional transmutation, all dimensionless parameters in the continuum limit are completely determined by the basic structure of the initial Lagrangian, without any continuous parameters to tune. In the limit of vanishing pion mass, the rho to nucleon mass ratio should be determined from first principles; it is the goal of lattice gauge theory to calculate just such numbers.

As we go below four dimensions, it is possible that additional couplings can become "relevant," requiring the renormalization-group picture of flow towards non-trivial fixed points. This is the basis of the renormalization group argument for "universality classes," often differing in basic symmetries.

One might imagine dimensionality as a continuously variable parameter. Then, just below four dimensions a renormalizable coupling becomes "super-renormalizable" and a non-trivial fixed point breaks away from vanishing coupling. Near four dimensions, this is at small coupling, forming the basis for an expansion in $4 - d$. This has become a major industry, making remarkably accurate predictions for critical exponents in three-dimensional systems [76].

An important consequence of this discussion is that a lattice action is in general highly non-unique. One can always add irrelevant operators and expect to obtain the same continuum limit.[3] Alternatively, one might hope to expedite the approach to the continuum limit by a judicious choice of the lattice action.

The renormalization group is indeed a rich subject. We have only touched on a few issues that are particularly valuable for the lattice theory. Perhaps the most remarkable result of this section is how a perturbative analysis of the renormalization-group equation gives rise to non-perturbative information on particle masses.

---

[3] One caveat is that the "irrelevent" operator might shift the flow sufficiently that it approaches a distinct fixed point. This is the case with the "Wilson term" for dynamical fermions, as will be discussed in Chapter 16.

Chapter 10

Classical gauge fields and topology

The renormalization group analysis demonstrates that non-perturbative effects are crucial to understanding the continuum limit of QCD on the lattice. However, the importance of going beyond the perturbation expansion was dramatically exposed from a completely different direction with the discovery of non-trivial classical solutions to the Yang-Mills theory. These are characterized by an explicit essential singularity at vanishing coupling. In this and the next chapter, we review these solutions and some rather interesting consequences for the Dirac operator.

Let us start with some basic definitions to establish notation in continuum language. Being ultimately interested in QCD, we concentrate on the gauge group $SU(N)$. This group has $N^2 - 1$ generators denoted λ^α. They are traceless N by N Hermitian matrices and satisfy the commutation relations

$$[\lambda^\alpha, \lambda^\beta] = if^{\alpha\beta\gamma}\lambda^\gamma, \tag{10.1}$$

involving the group structure constants $f^{\alpha\beta\gamma}$. By convention, these generators are orthogonalized and normalized:

$$\mathrm{Tr}\lambda^\alpha\lambda^\beta = \frac{1}{2}\delta^{\alpha\beta}. \tag{10.2}$$

For $SU(2)$ the generators are the spin matrices $\lambda^\alpha = \sigma^\alpha/2$, and the structure constants, the three indexed anti-symmetric tensor $f^{\alpha\beta\gamma} = \epsilon^{\alpha\beta\gamma}$.

Associated with each of the generators λ^α is a gauge potential $A_\mu^\alpha(x)$. For the classical theory, assume for now that these are differentiable functions of space-time and vanish rapidly at infinity.[1] The notation simplifies a bit by defining a matrix valued field A

$$A_\mu = A_\mu^\alpha \lambda^\alpha. \tag{10.3}$$

The covariant derivative is a matrix valued differential operator defined as

$$D_\mu = \partial_\mu + ig A_\mu. \tag{10.4}$$

Given the gauge potential, the corresponding matrix valued field strength is

$$F_{\mu\nu} = \frac{-i}{g}[D_\mu, D_\nu] = \partial_\mu A_\nu - \partial_\nu A_\mu + ig[A_\mu, A_\nu]$$
$$= D_\mu A_\nu - D_\nu A_\mu. \tag{10.5}$$

We define the dual field strength as

$$\tilde{F}_{\mu\nu} = \frac{1}{2}\epsilon_{\mu\nu\rho\sigma} F_{\rho\sigma}, \tag{10.6}$$

with $\epsilon_{\mu\nu\rho\sigma}$ being the anti-symmetric tensor with $\epsilon_{1234} = 1$.

In terms of the field strength, the classical Yang-Mills action is

$$S = \frac{1}{2}\int d^4x \; \mathrm{Tr}\; F_{\mu\nu} F_{\mu\nu}, \tag{10.7}$$

and the equations of motion are

$$D_\mu F_{\mu\nu} = 0. \tag{10.8}$$

This defines the classical Yang-Mills theory.

The Jacobi identity

$$[A, [B, C]] + [B, [C, A]] + [C, [A, B]] = 0 \tag{10.9}$$

applied to the covariant derivative implies that

$$\epsilon_{\mu\nu\rho\sigma} D_\nu F_{\rho\sigma} = 0. \tag{10.10}$$

Using $\tilde{F}_{\mu\nu} = \epsilon_{\mu\nu\rho\sigma} F_{\rho\sigma}$, this reads $D_\mu \tilde{F}_{\mu\nu} = 0$, and immediately implies that any self-dual or anti-self-dual field with $F = \pm\tilde{F}$ automatically satisfies the classical equations of motion. This is an interesting relation since $F = \tilde{F}$ is linear in derivatives of the gauge potential. This leads to a

---

[1] For the quantum theory, this assumption of differentiability is a subtle issue to which we will return.

multitude of known solutions [77]. Here we concentrate on just the simplest non-trivial one.

This theory is, after all, a gauge theory and therefore has a local symmetry. We previously discussed this in the lattice context, but historically it was originally motivated by the continuum gauge transformations of the classical theory, which we now review. Let $h(x)$ be a space-dependent element of $SU(N)$ in the fundamental representation. Assume that h is differentiable. Now define the gauge transformed field

$$A_\mu^{(h)} \to h^\dagger A_\mu h - \frac{i}{g} h^\dagger (\partial_\mu h). \tag{10.11}$$

This transformation takes a simple form for the covariant derivative

$$h^\dagger D_\mu h = D_\mu^{(h)} = \partial_\mu + ig A_\mu^{(h)}. \tag{10.12}$$

Similarly, for the field strength we have

$$F_{\mu\nu}^{(h)} = h^\dagger F_{\mu\nu} h. \tag{10.13}$$

Thus the action is invariant under this transformation, $S(A) = S(A^{(h)})$.

10.1. Surface terms

A remarkable feature of this formalism is that the combination Tr $F\tilde{F}$ is a total derivative. To see this, first construct

$$\begin{aligned} F\tilde{F} &= \frac{1}{2}\epsilon_{\mu\nu\rho\sigma}(2\partial_\mu A_\nu + ig A_\mu A_\nu)(2\partial_\rho A_\sigma + ig A_\rho A_\sigma) \\ &= \frac{1}{2}\epsilon_{\mu\nu\rho\sigma}\left(4\partial_\mu A_\nu \partial_\rho A_\sigma + 4ig(\partial_\mu A_\nu)A_\rho A_\sigma - g^2 A_\mu A_\nu A_\rho A_\sigma\right). \end{aligned} \tag{10.14}$$

If we take a trace of this quantity, the last term will drop out due to cyclicity. Thus,

$$\text{Tr}F\tilde{F} = 2\partial_\mu \epsilon_{\mu\nu\rho\sigma}\text{Tr}\left(A_\nu \partial_\rho A_\sigma + ig A_\nu A_\rho A_\sigma\right) = 2\partial_\mu K_\mu \tag{10.15}$$

where we define

$$K_\mu \equiv \epsilon_{\mu\nu\rho\sigma}\text{Tr}\left(A_\nu \partial_\rho A_\sigma + 2ig A_\nu A_\rho A_\sigma\right). \tag{10.16}$$

Note that although Tr $F\tilde{F}$ is gauge invariant, this is not true for K_μ.

Being a total derivative, the integral of this quantity

$$\int d^4x \,\frac{1}{2}\mathrm{Tr}F\tilde{F} \tag{10.17}$$

would vanish if we ignore surface terms. What is remarkable is that there exist finite action gauge configurations for which this does not vanish, even though the field strengths all go to zero rapidly at infinity. This is because the gauge fields A_μ that appear explicitly in the current K_μ need not necessarily vanish as rapidly as $F_{\mu\nu}$.

These surface terms are closely tied to the topology of the gauge potential at large distances. As we want the field strengths to go to zero at infinity, the potential should approach a pure gauge form $A_\mu \to \frac{-i}{g}h^\dagger\partial_\mu h$. In this case,

$$K_\mu \to -\frac{1}{g^2}\epsilon_{\mu\nu\rho\sigma}\mathrm{Tr}(h^\dagger\partial_\nu h)(h^\dagger\partial_\rho h)(h^\dagger\partial_\sigma h). \tag{10.18}$$

Note the similarity of this form to that for the group measure in Eq. (7.9). As with the measure, it is invariant if we take $h \to h'h$, with h' being an arbitrary fixed group element. The surface at infinity is topologically a three-dimensional sphere S_3. If we concentrate temporarily on $SU(2)$, this is the same as the topology of the group space. For larger groups, we can restrict h to a $SU(2)$ subgroup and proceed similarly. Thus, the integral of K_μ over the surface reduces to the integral of h over a sphere with the invariant group measure. This can give a non-vanishing contribution if the mapping of h onto the sphere at infinity covers the entire group in a non-trivial manner. Mathematically, one can map the S_3 of spatial infinity onto the S_3 of group space an integral number of times, i.e. $\Pi_3(SU(2)) = Z$. Thus we have

$$\int d^4x \,\frac{1}{2}\mathrm{Tr}F\tilde{F} \propto \nu, \tag{10.19}$$

where ν is an integer describing the number of times $h(x)$ wraps around the group as x covers the sphere at infinity. The normalization involves the surface area of a three-dimensional sphere and can be worked out with the result

$$\int d^4x \,\frac{1}{2}\mathrm{Tr}F\tilde{F} = \frac{8\pi^2\nu}{g^2}. \tag{10.20}$$

For groups larger than $SU(2)$, one can smoothly deform h to lie in an $SU(2)$ subgroup, and thus this quantization of the surface term applies to any $SU(N)$.

If we were to place such a configuration into the path integral for the quantum theory, we might expect a suppression of these effects by a factor of $\exp(-8\pi^2/g^2)$. This is clearly non-perturbative; however, this factor strongly underestimates the importance of topological effects. The problem is that we only need to excite non-trivial fields over the quantum mechanical vacuum, not the classical one. The correct suppression is indeed exponential in the inverse coupling squared, but the coefficient in the exponent can be determined from asymptotic freedom and dimensional transmutation. We will return to this point in Chapter 14.

The combination $\text{Tr}F\tilde{F}$ is formally an operator of dimension four, the same as the basic gauge theory action density $\text{Tr}FF$. This naturally leads to the question of what would happen if we consider a new action which also includes this parity-odd term. Classically, it does nothing since it reduces to the surface term described above. However, quantum mechanically this is no longer the case. As we will discuss extensively later, the physics of QCD depends quite non-trivially on such a term. An interesting feature follows from the quantization of the resulting surface term. Because of the above quantization and an imaginary factor in the path integral, physics is periodic in the coefficient of $\frac{1}{2}\text{Tr}F\tilde{F}$. Although discussing the consequences directly with such a term in the action is traditional, we will follow a somewhat different path in later sections and introduce this physics through its effects on fermions.

10.2. An explicit solution

To demonstrate that non-trivial winding solutions indeed exist, we specialize to $SU(2)$ and find an explicit example. To start, consider the positivity of the norm of $F \pm \tilde{F}$:

$$0 \le \int d^4x \ (F \pm \tilde{F})^2 = 2 \int d^4x \ F^2 \pm 2 \int d^4x \ F\tilde{F}. \tag{10.21}$$

This means that the action is bounded below by the absolute value of $\int d^4x \ \frac{1}{2}\text{Tr}F\tilde{F}$. This bound is reached only if $F = \pm\tilde{F}$. As mentioned earlier, reaching this is sufficient to guarantee a solution to the equations of motion. We will now explicitly construct such a self-dual configuration.

Start with a gauge transformation function which is singular at the origin but maps around the group at a constant radius:

$$h(x_\mu) = \frac{t + i\vec{\tau} \cdot \vec{x}}{\sqrt{x^2}} = \frac{T_\mu x_\mu}{|x|}. \tag{10.22}$$

Here we define the four-component object $T_\mu = \{1, i\vec{\tau}\}$. Considering space with the origin removed, construct the pure gauge field

$$B_\mu = \frac{-i}{g} h^\dagger \partial_\mu h = \frac{-i}{g} h^\dagger (T_\mu x^2 - x_\mu T \cdot x)/|x|^3. \qquad (10.23)$$

Because this is nothing but a gauge transformation of a vanishing gauge field, the corresponding field strength automatically vanishes, i.e.

$$\partial_\mu B_\nu - \partial_\nu B_\mu + ig[B_\mu, B_\nu] = 0. \qquad (10.24)$$

This construction gives a unit winding at infinity. However, this gauge field is singular at the origin where the winding unwraps. If we smooth this singularity at $x = 0$ with a field of form

$$A_\mu = f(x^2) B_\mu \qquad (10.25)$$

where $f(0) = 0$ and $f(\infty) = 1$, this will remove the unwrapping at the origin and automatically leave a field configuration with non-trivial winding. The idea is to find a particular $f(x^2)$ such that A also gives a self-dual-field strength and thereby is a solution to the equations of motion.

We have set things up symmetrically under space-time rotations about the origin. This connection with $O(4)$ is convenient in that we only need to verify the self duality along a single direction. Consider this to be the time axis, along which self duality requires

$$F_{01}(\vec{x} = 0, t) = \pm F_{23}(\vec{x} = 0, t). \qquad (10.26)$$

A little algebra gives

$$F_{\mu\nu} = (f - f^2)(\partial_\mu B_\nu - \partial_\nu B_\mu) + 2f'(x_\mu B_\nu - x_\nu B_\mu), \qquad (10.27)$$

so along the time axis we have

$$F_{01} \rightarrow 2f't\frac{\tau_1}{gt} \qquad F_{23} \rightarrow (f - f^2)\frac{2\tau_1}{gt^2}. \qquad (10.28)$$

Thus, the self duality condition reduces to a simple first order differential equation

$$z f'(z) = \pm(f - f^2). \qquad (10.29)$$

This is easily solved to give

$$f(z) = \frac{1}{1 + \rho^2 z^{\pm 1}} \qquad (10.30)$$

where ρ is an arbitrary constant of integration. To have the function vanish at the origin, we take the minus solution. The resulting form for the gauge field

$$A_\mu = \frac{x^2}{x^2 + \rho^2} B_\mu = \frac{-ix^2}{g(x^2 + \rho^2)} h^\dagger \partial_\mu h \tag{10.31}$$

is the self-dual instanton. The parameter ρ controls the size of the configuration. Its arbitrary value is a consequence of the conformal invariance of the classical theory. Switching h and $h^\dagger$ gives a solution with the opposite winding.

This solution is characterized by size ρ and location in four-dimensional space-time. And, of course, any gauge transformation on one solution gives another. These field configurations have been called "instantons" or "pseudo-particles." This distinguishes them from conventional "particles," which follow a world line through time. In the next chapter, we turn to some remarkable features of the Dirac equation in the presence of such a solution.

Chapter 11

Zero modes and the Dirac operator

A particularly important and intriguing aspect of these classical field configurations is that they support exact zero modes for the Dirac operator. We will shortly discuss the rigorous connection between the gauge field winding and the zero modes of the Dirac operator. Here, we verify this connection explicitly for the above solution. Thus, we look for a spinor field $\psi(x)$ satisfying

$$\gamma_\mu D_\mu \psi(x) = \gamma_\mu \left(\partial_\mu + ig A_\mu \right) \psi(x) = 0 \tag{11.1}$$

where we insert the gauge field from Eq. (10.31). The wave function ψ is a spinor in Dirac space and a doublet in $SU(2)$ space; i.e. it has 8 components. Similarly, $\gamma_\mu A_\mu$ is an 8 by 8 matrix, with a factor of four from spinor space and a factor of two from the internal gauge symmetry. The solution entangles all of these indices in a non-trivial manner.

Since we do not want a singularity in ψ at the origin, it is natural to look for a solution of the form

$$\psi(x) = p(|x|)V, \tag{11.2}$$

where p is a scalar function of the four-dimensional radius and V is a constant vector in spinor and color space. As before, it is convenient to look for the solution along the time axis. There, A_0 vanishes and we have

$$\vec{A} = \frac{1}{g} \frac{t}{t^2 + \rho^2} \vec{\tau}. \tag{11.3}$$

89

Then the equation of interest reduces to

$$\gamma_0 \frac{d}{dt} \psi(t) = -\frac{t}{t^2 + \rho^2} \, \vec{\tau} \cdot \vec{\gamma} \, \psi(t). \tag{11.4}$$

The 8 by 8 matrix $\vec{\tau} \cdot \vec{\gamma}$ is readily diagonalized, giving the eigenvalues

$$\lambda \in \{-3, -1, -1, -1, 1, 1, 1, 3\}. \tag{11.5}$$

Only the $+3$ eigenvalue gives a normalizable solution

$$\psi(t) = \psi(0) \exp\left(-3 \int_0^t \frac{t \, dt}{t^2 + \rho^2}\right). \tag{11.6}$$

For general x_μ, this becomes

$$\psi(x) = \psi(0) \left(\frac{\rho^2}{x^2 + \rho^2}\right)^{3/2}. \tag{11.7}$$

At large x, this goes at x^{-3} so its square is normalizable. None of the other eigenvalues of $\vec{\tau} \cdot \vec{\gamma}$ give a normalizable wave function; thus, the solution is unique.

We see the appearance of a direct product of two $SU(2)$'s, one from spin and one from isospin. For the large eigenvalue, these combine as an overall singlet. The other positive eigenvalues of $\vec{\tau} \cdot \vec{\gamma}$ represent the triplet combination, while the negative eigenvalues come from antiparticle states.

This zero eigenvalue of D is robust under smooth deformations of the gauge field. This is because the anti-commutation of D with γ_5 says that all non-zero eigenvalues of D occur in conjugate pairs. Without bringing in another eigenvalue, the isolated one at zero cannot move. In the next section we demonstrate the general result that, for arbitrary smooth gauge fields, the number of zero modes of the Dirac operator is robustly given by the topological winding number.

11.1. The index theorem

We have demonstrated a zero action solution to the Dirac equation associated with a particular topologically non-trivial gauge configuration. Here we generalize this result, deriving the index theorem relating zero modes of the Dirac operator to the overall topology of the gauge field. We work directly with the naive continuum Dirac operator. Assume for this section that the gauge fields are smooth and differentiable. As pointed out in Chapter 5, this is suspect for typical fields in the path integral. Here our goal is only to show that robust zero modes must exist already in the classical theory. Later we will see that the generalization of these zero modes to the quantum theory is intimately tied to certain quantum mechanical anomalies crucial to non-perturbative physics.

The combination of the anti-Hermitian character of the classical Dirac operator $D = \gamma_\mu D_\mu$ along with its anti-commutation with γ_5 shows that the non-zero eigenvalues of D all occur in complex conjugate pairs. In particular, if we have

$$D|\psi\rangle = \lambda|\psi\rangle, \tag{11.8}$$

then we immediately obtain the conjugate eigenvector from

$$D\gamma_5 |\psi\rangle = -\lambda \gamma_5|\psi\rangle. \tag{11.9}$$

Since $|\psi\rangle$ and $\gamma_5|\psi\rangle$ have different eigenvalues under the anti-Hermitian operator D, they must be orthogonal

$$\langle\psi|\gamma_5|\psi\rangle = 0. \tag{11.10}$$

On the other hand, any exact zero eigenmodes need not be paired. Furthermore, when restricted to the space of zero eigenmodes, γ_5 and D commute and can be simultaneously diagonalized. The eigenvalues of γ_5 are all either plus or minus unity. Combining all these ideas together gives a simple method to count the number of zero modes of the Dirac operator weighted by their chirality. In particular, we have the relation

$$\nu = n_+ - n_- = \text{Tr}\gamma_5 e^{D^2/\Lambda^2} \tag{11.11}$$

where $n_\pm$ denotes the number of zero modes with eigenvalue ± 1 under γ_5. Here, the parameter Λ is introduced to control the behavior of the trace as the eigenvalues go to infinity. It can be thought of as a regulator, although the above equation is independent of its value. The remarkable result is that in a given gauge field, this integer is identical to the winding number from Eq. (10.20).

To proceed, we first write the square of the Dirac operator appearing in the above exponential:

$$D^2 = \partial^2 - g^2 A^2 + 2igA_\mu\partial_\mu + ig(\partial_\mu A_\mu) - \frac{g}{2}\sigma_{\mu\nu}F_{\mu\nu}, \tag{11.12}$$

where $[\gamma_\mu, \gamma_\nu] = 2i\sigma_{\mu\nu}$. Expanding Eq. (11.11) for the winding number in powers of the gauge field, the first non-vanishing term appears in the fourth power of the Dirac operator. This involves two powers of the sigma matrices through the relation

$$\text{Tr } \gamma_5 \sigma_{\mu\nu} \sigma_{\rho\sigma} = 4\epsilon_{\mu\nu\rho\sigma}. \tag{11.13}$$

Thus, our expression for the winding number becomes

$$\nu = \text{Tr} \gamma_5 e^{D^2/\Lambda^2} = \frac{g^2}{2\Lambda^4} \text{Tr}_{x,c} e^{\partial^2/\Lambda^2} \epsilon_{\mu\nu\rho\sigma} F_{\mu\nu} F_{\rho\sigma} + O(\Lambda^{-6}) \tag{11.14}$$

where $\text{Tr}_{x,c}$ refers to the trace over space and color; the trace over the spinor index gives rise to the anti-symmetric tensor factor. We will now show how the trace over the space index gives a divergent factor that cancels the Λ^{-4} prefactor. Higher order terms go to zero rapidly enough with Λ to be ignored.

The factor e^{∂^2/Λ^2} serves to mollify traces over position space. Consider some function $f(x)$ as representing a diagonal matrix in position space $M(x, x') = f(x)\delta(x - x')$. The formal trace would be $\text{Tr} M = \int dx M(x, x)$, but this diverges since it involves a delta function of zero. Writing the delta function in terms of its Fourier transform,

$$e^{\partial^2/\Lambda^2} \delta(x - x') = \int \frac{d^4 p}{(2\pi)^4} e^{ip\cdot(x-x')} e^{-p^2/\Lambda^2}$$

$$= \frac{\Lambda^4}{16\pi^2} e^{-(x-x')^2 \Lambda^2/4} \tag{11.15}$$

shows how this "heat kernel" spreads the delta function. This regulates the desired trace

$$\text{Tr}_x f(x) \equiv \frac{\Lambda^4}{16\pi^2} \int d^4 x f(x). \tag{11.16}$$

Using this to remove the spatial trace in the above gives the well-known relation:

$$\nu = \frac{g^2}{32\pi^2} \text{Tr}_c \int d^4 x \epsilon_{\mu\nu\rho\sigma} F_{\mu\nu} F_{\rho\sigma} = \frac{g^2}{16\pi^2} \text{Tr}_c \int d^4 x F_{\mu\nu} \tilde{F}_{\mu\nu}. \tag{11.17}$$

As discussed earlier, this involves a total derivative that partially integrates into an expression over spatial infinity and counts the topological winding of the gauge field. Thus, the number of zero modes of the Dirac operator is an equivalent way to determine this topology. The index theorem represents the fact that Eqs. (10.20) and (11.17) have identical content despite rather different derivations.

11.2. Topology and eigenvalue flow

There is a close connection between the zero modes of the Dirac operator in the Euclidean path integral and a flow of eigenvalues of the fermion Hamiltonian in Minkowski space. To see how this works, it is convenient to work in the temporal gauge with $A_0 = 0$ and separate out the space-like part of the Dirac operator

$$D = \gamma_0(\partial_0 + H(\vec{A}(t))). \tag{11.18}$$

Here, H involves space derivatives and gamma matrices. Consider $\vec{A}$ as a time dependent gauge field through which the fermions propagate. Assume that at large positive or negative times this background field becomes constant in time. Without a mass term, the continuum theory H commutes with γ_5 and anti-commutes with γ_0. Therefore, its eigenvalues appear in pairs of opposite energy and opposite chirality; *i.e.* if we have

$$H\phi = E\phi,$$
$$\gamma_5\phi = \pm\phi, \tag{11.19}$$

then

$$H\gamma_0\phi = -E\gamma_0\phi,$$
$$\gamma_5\gamma_0\phi = \mp\gamma_0\phi. \tag{11.20}$$

Now, suppose we diagonalize H at some given time

$$H(\vec{A}(t))\phi_i(t) = E_i(t)\phi_i(t) \tag{11.21}$$

where the wave function $\phi(t)$ implicitly depends on space, spinor, and color indices. Suppose, further, that we can find some eigenvalue that changes adiabatically from negative to positive in going from large negative to large positive time, as sketched in Fig. 11.1. From this particular eigenstate, construct the four-dimensional field

$$\psi(t) = e^{-\int_0^t E(t')\, dt'}\phi(t). \tag{11.22}$$

Because of the change in sign of the energy, the exponential factor function goes to zero at both positive and negative large times, as sketched in Fig. 11.2.

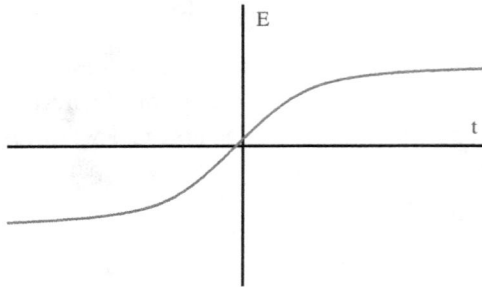

Figure 11.1:   An energy eigenvalue that changes in sign between the distant past and future.

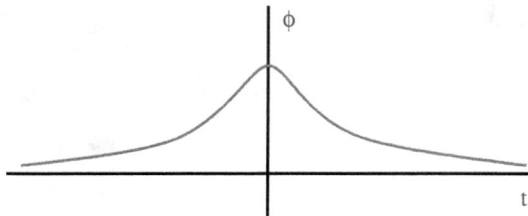

Figure 11.2:   The adiabatic evolution gives rise to a normalizable zero mode of the four-dimensional Dirac operator.

If we now consider the four-dimensional Dirac operator applied to this function, we obtain

$$(\gamma_0 \partial_0 + \gamma_0 H(\vec{A}(t)))\psi(t) = O(\partial_0 \psi(t)). \qquad (11.23)$$

If the evolution is adiabatic, the last term is small and we have an approximate zero mode.

The assumption of adiabaticity is unnecessary in the chiral limit of zero mass. Then, the eigenvalues of D are either real or occur in complex conjugate pairs. Any unaccompanied eigenvalue of the four-dimensional D occurs robustly at zero. This is another manifestation of the index theorem; we can count Euclidean-space zero modes by studying the zero crossings appearing in the eigenvalues of the Minkowski-space Hamiltonian.

In the above construction, the evolving eigenmode of H is accompanied by another of opposite energy and chirality. Inserted into Eq. (11.22), this will give a non-normalizable form for the four-dimensional field. Thus, we obtain only a single normalizable zero mode for the Euclidean Dirac operator. Note that if a small mass term is included, the up-going and down-going

Hamiltonian eigenstates will mix and the crossing is forbidden. In this case, the mass term shifts the modes of D away from zero, although, when they are unpaired, they robustly remain real.

This eigenvalue flow provides an intuitive picture of the anomaly [78]. Start at early times with a filled Dirac sea and all negative-energy eigenstates filled, and then slowly evolve through one of the above crossings. In the process, one of the filled states moves to positive energy, leaving a non-empty positive energy state. At the same time, the opposite chirality state moves from positive to negative energy. As long as the process is adiabatic, we wind up at large time with one filled positive-energy state and one empty negative-energy state. As these are of opposite chirality, effectively chirality is not conserved. The result is particularly dramatic in the weak interactions, where anomalies are canceled between quarks and leptons. This flow from negative to positive energy states theoretically results in baryon non-conservation, although at a rate too small to have been observed [26].

Chapter 12

Chiral symmetry

Chirality refers to the parity symmetry under a mirror reflection. The name comes from Lord Kelvin [79] and is most frequently used in chemistry, where many molecules come in chiral pairs. These have the same chemical formula, but are transformed non-trivially into each other under parity. In particle physics, this symmetry is related to how massless particles are special under Lorentz transformations. If a particle moves at the speed of light, its spin along the direction of motion is frame invariant. Basically, one cannot overtake something moving at the speed of light.

Much older a tool than the lattice, ideas based on chiral symmetry have historically provided considerable insight into how the strong interactions work. In particular, chirality is crucial to our understanding of why the pion is so much lighter than the rho, despite them both being made of the same quarks. Combining these ideas with the lattice has provided considerable insight into many non-perturbative issues in QCD. Here, we review the basic ideas of chiral symmetry for the strong interactions. A crucial aspect of this discussion is the famous anomaly and its consequences for the η' meson.

The classical Lagrangian for QCD couples left and right handed quark fields only through mass terms. If we project out the two helicity states of a quark field by defining

$$\psi_L = \frac{1 + \gamma_5}{2}\psi \qquad \psi_R = \frac{1 - \gamma_5}{2}\psi, \tag{12.1}$$

then the classical kinetic term for the quarks separates, so that

$$i\overline{\psi}D\psi = i\overline{\psi}_L D\psi_L + i\overline{\psi}_R D\psi_R. \tag{12.2}$$

On the other hand, the mass term directly mixes the two chiralities:

$$m\overline{\psi}\psi = m\overline{\psi}_R\psi_L + m\overline{\psi}_L\psi_R. \qquad (12.3)$$

Thus, naively, the massless theory appears to have independent conserved currents associated with each handedness. After confinement, it is generally understood that the axial chiral symmetry is spontaneously broken via the scalar field $\sigma \propto \overline{\psi}\psi$ acquiring a vacuum expectation value. As with the mass term, this field mixes chiralities. This expectation value is not invariant under the axial chiral symmetry. Although chiral symmetry is broken, parity is not; it is only the independence of the left and right quark fields that the vacuum does not respect. The consequences of this spontaneous breaking are deep, and are the topic of much of the following material.

For N_f massless flavors, there is classically an independent $U(N_f)$ symmetry associated with each chirality, giving what is often written in terms of axial and vector fields as a $U(N_f)_V \otimes U(N_f)_A$ symmetry. As is well-known, this full symmetry does not survive quantization, being broken to a $SU(N_f)_V \otimes SU(N_f)_A \otimes U(1)_B$, where the $U(1)_B$ represents the symmetry of baryon number conservation. The only surviving axial symmetries of the massless quantum theory are non-singlet under flavor symmetry.

As we will see, breaking of the classical $U(1)$ axial symmetry is closely tied to the possibility of introducing, into massive QCD, a CP-violating parameter, usually called Θ. For an extensive review, see Ref. [80]. While such a term is allowed from fundamental principles, experimentally it appears to be extremely small. This raises an unresolved puzzle for attempts to unify the strong interactions with the weak. Since the weak interactions do violate CP, why is there no residue of this remaining in the strong sector below the unification scale?

One goal of this chapter is to provide a qualitative understanding of the role of the Θ parameter in meson physics. We concentrate on symmetry alone and do not attempt to rely on any specific form for an effective Lagrangian. We build on the connection between Θ and a flavor-singlet Z_{N_f} symmetry that survives the anomaly. We will see that, when the lightest quarks are made massive and degenerate, a first order transition must occur when Θ passes through π. This transition is quite generic, but can be avoided under limited conditions with one quark considerably lighter than the others. This discussion should also make it clear that the sign of the quark mass is physically relevant for an odd number of flavors. This is a non-perturbative effect that is invisible to naive diagrammatic treatments.

Throughout this chapter, we use the language of continuum field theory. Of course, underlying this we must assume some non-perturbative regulator has been imposed so that we can make sense of various products of fields, such as the condensing combination $\sigma = \bar{\psi}\psi$. For a momentum space cutoff, assume that it is much larger than Λ_{QCD}. Correspondingly, for a lattice cutoff imagine that the lattice spacing is much smaller than $1/\Lambda_{QCD}$. In this chapter, we ignore any lattice artifacts that should vanish in the continuum limit. We will return to such issues later when we discuss lattice fermions.

12.1. Effective potentials

To formally explore how a scalar field might acquire a vacuum expectation value, we review the concept of effective potentials in quantum field theory. In generic continuum field theory language, consider the path integral for a scalar field

$$Z = \int d\phi \, e^{-S(\phi)}. \tag{12.4}$$

After adding in some external sources, denoted by J,

$$Z(J) = \int d\phi \, e^{-S(\phi)+J\phi}, \tag{12.5}$$

general correlation functions can be found by differentiating with respect to J. Here, we use a shorthand notation that suppresses the space dependence; *i.e.* $J\phi = \int dx J(x)\phi(x)$ in the continuum, or $J\phi = \sum_i J_i\phi_i$ on a lattice. For this discussion, it is unimportant whether the field ϕ is "fundamental" in the sense of appearing directly in a Lagrangian, or is some composite field that can create a physical particle of interest.

One can think of J as an external force pulling on the field. Such a force will tend to drive the field to have an expectation value

$$\langle\phi\rangle_J = -\frac{\partial F}{\partial J} \tag{12.6}$$

where the free energy in the presence of the source is defined as $F(J) = -\log(Z(J))$.

Now imagine inverting Eq. (12.6) to determine what value of the force J would be needed to give some desired expectation value Φ; *i.e.* we want to solve

$$\Phi(J) = \langle\phi\rangle_{J(\Phi)} = -\frac{\partial F}{\partial J} \tag{12.7}$$

for $J(\Phi)$. In terms of this formal solution, construct the "Legendre transform"

$$V(\Phi) = F(J(\Phi)) + \Phi J(\Phi) \tag{12.8}$$

and look at

$$\frac{\partial V}{\partial \Phi} = -\Phi \frac{\partial J}{\partial \Phi} + J + \Phi \frac{\partial J}{\partial \Phi} = J. \tag{12.9}$$

If we now turn off the sources, this derivative vanishes. Thus, the expectation value of the field in the absence of sources occurs at an extremum of $V(\Phi)$. This quantity V is referred to as the "effective potential."

An interesting formal property of this construction follows from looking at the second derivative of V,

$$\frac{\partial^2 V}{\partial \Phi^2} = \frac{\partial J}{\partial \Phi}. \tag{12.10}$$

Actually, it is easier to look at the inverse

$$\frac{\partial \Phi}{\partial J} = -\frac{\partial^2 F}{\partial J^2} = \langle \phi^2 \rangle - \langle \phi \rangle^2 = \langle (\phi - \langle \phi \rangle)^2 \rangle \geq 0. \tag{12.11}$$

Thus, this second derivative is never negative! This shows we are actually looking for a minimum and not a maximum of V. It also implies that $V(\Phi)$ can only have ONE minimum!

This convexity property is usually ignored in conventional discussions, where phase transitions are signaled by jumps between distinct minima of the potential. So what is going on? Are phase transitions impossible? Physically, the more you pull on the field, the larger the expectation of Φ will become. It will not go back. The proper interpretation is that we must do Maxwell's construction. If we force the expectation of ϕ to lie between two distinct stable phases, the system will separate into a heterogeneous mixture of these phases. In this region, the effective potential is flat. Note that there is no large volume limit required in the above discussion. However, other definitions of V can allow a small barrier at finite volume due to surface tension effects. A mixed phase must contain interfaces, and their energy represents a small barrier.

12.2. Goldstone Bosons

Now we turn to a brief discussion on some formal aspects of Goldstone bosons. Suppose we have a field theory containing a conserved current

$$\partial_\mu j_\mu = 0 \tag{12.12}$$

so the corresponding charge $Q = \int d^3x j_0(x)$ is a constant

$$\frac{dQ}{dt} = -i[H, Q] = 0. \tag{12.13}$$

Here, H is the Hamiltonian for the system under consideration. Suppose, however, that for some reason the vacuum is not a singlet under this charge

$$Q|0\rangle \neq 0. \tag{12.14}$$

Then there must exist a massless particle in the theory. Consider the state

$$\exp(i\theta \int d^3x j_0(x) e^{-\epsilon x^2})|0\rangle \tag{12.15}$$

where ϵ is a convenient cutoff, and θ, some parameter. As epsilon goes to zero, this state, by assumption, is not the vacuum. However, since the Hamiltonian commutes with Q, the expectation value of the Hamiltonian in this state goes to zero with ϵ.[1] We can thus find a state that is not the vacuum but with arbitrarily small energy. The theory has no mass gap. This situation of having a symmetry under which the vacuum is not invariant is referred to as "spontaneous symmetry breaking." The low energy states represent massless particles called Goldstone bosons [81].

Free massless field theory is a marvelous example where everything can be worked out. The massless equation of motion

$$\partial_\mu \partial_\mu \phi = 0 \tag{12.16}$$

can be written in the form

$$\partial_\mu j_\mu = 0 \tag{12.17}$$

where

$$j_\mu = \partial_\mu \phi. \tag{12.18}$$

---

[1]Normalize the ground state energy to be zero.

The broken symmetry is the invariance of the Lagrangian $L = \int d^4x$ $(\partial_\mu \phi)^2/2$ under constant shifts of the field

$$\phi \to \phi + c. \tag{12.19}$$

Note that $j_0 = \partial_0 \phi = \pi$ is the conjugate variable to ϕ. Since it is a free theory, one could work out explicitly

$$\langle 0 | \exp(i\theta \int d^3x j_0(x) e^{-\epsilon x^2/2}) | 0 \rangle. \tag{12.20}$$

We can, however, save ourselves the work using dimensional analysis. The field ϕ has dimensions of inverse length, while j_0 goes as inverse squared length. Thus, θ above has units of inverse length. These are the same dimensions as ϵ^2. Now for a free theory, by Wick's theorem, the answer must be Gaussian in θ. We conclude that the above overlap must go as

$$\exp(-C\theta^2/\epsilon^4) \tag{12.21}$$

where C is some non-vanishing dimensionless number. This expression rapidly goes to zero as epsilon becomes small, showing that the vacuum is indeed not invariant under the symmetry. In the limit of ϵ going to zero, we obtain a new vacuum that is not even in the same Hilbert space. The overlap of this new state with any local polynomial of fields on the original vacuum vanishes.

It is perhaps interesting to note that the canonical commutation relations $[\pi(x), \phi(y)] = i\delta(x - y)$ imply, for the currents,

$$[j_0(x), j_i(y)] = -i\frac{d}{dx}\delta(x - y). \tag{12.22}$$

In a Hamiltonian formulation, equal time commutators of different current components must involve derivatives of delta functions. This is a generic property and does not depend on the symmetry being spontaneously broken [82].

12.3.   Pions and spontaneous symmetry breaking

We now extend the effective potential to a function of several relevant meson fields in QCD. Intuitively, V represents the energy of the lowest state for a given field expectation, as discussed more formally earlier via a Legendre transformation. Here, we will ignore the result that effective potentials must be convex functions of their arguments. As discussed, this issue is easily understood in terms of a Maxwell construction involving the phase separation that will occur if one asks for a field expectation in what would otherwise be a concave region. Thus, we will use the traditional language of spontaneous symmetry breaking corresponding to having an effective potential with more than one minimum. When the underlying theory possesses some symmetry but the individual minima do not, spontaneous breaking comes about when the vacuum selects one of the minima arbitrarily. The discussion here closely follows that in Ref. [83].

We work here with the composite scalar and pseudo-scalar fields

$$\sigma \sim \overline{\psi}\psi,$$
$$\pi_\alpha \sim i\overline{\psi}\lambda_\alpha\gamma_5\psi, \tag{12.23}$$
$$\eta' \sim i\overline{\psi}\gamma_5\psi.$$

Here, the λ_α are the generators for the flavor group $SU(N_f)$. They are generalizations of the usual Gell-Mann matrices from $SU(3)$; however, now we are concerned with the flavor group, not the internal symmetry group related to confinement. As mentioned earlier, we must assume that some sort of regulator, perhaps a lattice, is in place to define these products of fields at the same point. Indeed, most of the quantities mentioned in this section are formally divergent, although we will concentrate on those aspects that survive the continuum limit.

To simplify the discussion, consider degenerate quarks with a small common mass m. Later, we will work out in some detail the two-flavor case for non-degenerate quarks. It is also convenient to initially restrict N_f to be even, saving, for later, some interesting subtleties arising with an odd number of flavors. And we assume N_f is small enough to maintain asymptotic freedom as well as avoiding any possible conformal phases.

The conventional picture of spontaneous chiral symmetry breaking in the limit of massless quarks assumes that the vacuum acquires a quark condensate with

$$\langle\overline{\psi}\psi\rangle = \langle\sigma\rangle = v \neq 0. \tag{12.24}$$

In terms of the effective potential, $V(\sigma)$ should acquire a double-well structure, as sketched in Fig. 12.1. The symmetry under $\sigma \leftrightarrow -\sigma$ is associated with the invariance of the action under a flavored chiral rotation. For example, with two flavors, the change of variables

$$\psi \to e^{i\pi\tau_3\gamma_5/2}\psi = i\tau_3\gamma_5\psi,$$
$$\overline{\psi} \to \overline{\psi}e^{i\pi\tau_3\gamma_5/2} = \overline{\psi}\, i\tau_3\gamma_5 \tag{12.25}$$

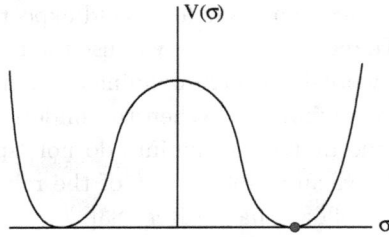

Figure 12.1:   Spontaneous chiral symmetry breaking is represented by a double-well effective potential with the vacuum settling into one of two possible minima. In this minimum, chiral symmetry is broken by the selection of a specific value for the quark condensate.

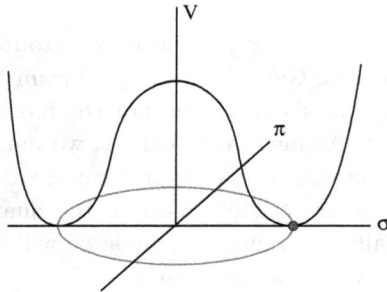

Figure 12.2:   The flavor non-singlet pseudo-scalar mesons are Goldstone bosons corresponding to flat directions in the effective potential. This shape is sometimes referred to as a "mexican hat" or "bottom of a wine bottle" potential.

leaves the massless action invariant but takes σ to its negative. Here, τ_3 is the conventional Pauli matrix corresponding to the third component of isospin.

Extending the effective potential to a function of the non-singlet pseudo-scalar fields gives the standard picture of Goldstone bosons. These are massless when the quark mass vanishes, corresponding to $N_f^2 - 1$ "flat" directions for the potential. One such direction is sketched schematically in Fig. 12.2. For the two-flavor case, these rotations represent a symmetry mixing the sigma field with the pions

$$
\begin{aligned}
\sigma &\rightarrow \quad \sigma \cos(\phi) + \pi^\alpha \sin(\phi), \\
\pi^\alpha &\rightarrow -\sigma \sin(\phi) + \pi^\alpha \cos(\phi).
\end{aligned}
\tag{12.26}
$$

In some sense, the pions are waves propagating through the non-vanishing sigma condensate. The oscillations of these waves occur in a direction

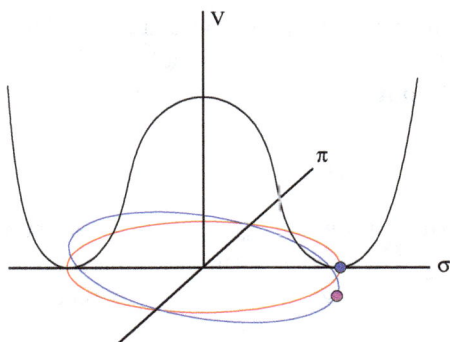

Figure 12.3: A small quark mass term tilts the effective potential, selecting one direction for the true vacuum and giving the Goldstone bosons a mass proportional to the square root of the quark mass.

"transverse" to the sigma expectation. They are massless because there is no restoring force in that direction.

If we now introduce a small mass for the quarks, this will effectively tilt the potential $V(\sigma) \rightarrow V(\sigma) - m\sigma$. This selects one minimum as the true vacuum. The tilting of the potential breaks the global symmetry and gives the Goldstone bosons a small mass proportional to the square root of the quark mass, as sketched in Fig. 12.3. The standard chiral Lagrangian approach is a simultaneous expansion in the masses and momenta of these light particles.

As discussed earlier, in a Hamiltonian approach, Goldstone bosons are associated with conserved currents with charges that do not leave the vacuum invariant. In the present case, these are the axial currents formally given by the quark bilinears

$$A_\mu^\alpha = \overline{\psi}\lambda^\alpha\gamma_\mu\gamma_5\psi. \tag{12.27}$$

Combined with the vector fields

$$V_\mu^\alpha = \overline{\psi}\lambda^\alpha\gamma_\mu\psi, \tag{12.28}$$

these give rise to the famous algebra of currents

$$[V_0^\alpha(x), V_0^\beta(y)] = if^{\alpha\beta\gamma}V_0^\gamma(x)\delta(x - y),$$
$$[V_0^\alpha(x), A_0^\beta(y)] = if^{\alpha\beta\gamma}A_0^\gamma(x)\delta(x - y),$$
$$[A_0^\alpha(x), A_0^\beta(y)] = if^{\alpha\beta\gamma}V_0^\gamma(x)\delta(x - y), \tag{12.29}$$

with the $f^{\alpha\beta\gamma}$ being the structure constants for the internal symmetry group. Indeed, it was this algebra that motivated Bjorken to propose the idea of scaling in deep inelastic lepton scattering [84, 85].

12.4. The Sigma model

Much of the structure of low energy QCD is nicely summarized in terms of an effective chiral Lagrangian formulated in terms of a field which is an element of the underlying flavor group. In this section, we review this model for the strong interactions with three quarks, namely up, down, and strange. The theory has an approximate SU(3) symmetry, broken by unequal masses for the quarks. We work with the familiar octet of light pseudo-scalar mesons π_α with $\alpha = 1\dots 8$ and consider an SU(3) valued field

$$\Sigma = \exp(i\pi_\alpha \lambda_\alpha / f_\pi) \in SU(3). \tag{12.30}$$

Here, the λ_α are the usual Gell-Mann matrices which generate the flavor group and f_π is a dimensional constant with a phenomenological value of about 93 MeV. We follow the normalization convention that $\mathrm{Tr}\lambda_\alpha\lambda_\beta = 2\delta_{\alpha\beta}$. The neutral pion and the eta meson will play a special role later in this review; they are the coefficients of the commuting generators

$$\lambda_3 = \frac{1}{\sqrt{3}} \begin{pmatrix} 1 & 0 & 0 \\ 0 & -1 & 0 \\ 0 & 0 & 0 \end{pmatrix} \tag{12.31}$$

and

$$\lambda_8 = \frac{1}{\sqrt{3}} \begin{pmatrix} 1 & 0 & 0 \\ 0 & 1 & 0 \\ 0 & 0 & -2 \end{pmatrix}, \tag{12.32}$$

respectively. In the chiral limit of vanishing quark masses, the interactions of the eight massless Goldstone bosons are modeled with the effective Lagrangian density

$$L_0 = \frac{f_\pi^2}{4} \mathrm{Tr}(\partial_\mu \Sigma^\dagger \partial_\mu \Sigma). \tag{12.33}$$

Expanding Eq. (12.33) to second order in the meson fields gives the conventional kinetic terms for our eight mesons.

    The non-linear constraint of Σ onto the group SU(3) makes this theory non-renormalizable. It is to be understood only as the starting point for an expansion of particle interactions in powers of momenta. This restriction of the fields to group elements also allows rather interesting topological interactions to appear at higher orders in momenta [86, 87]. This is a very rich subject and has some connections with fermion doubling [88]. However, most of the later discussions in this book do not impose this non-linear restriction, and for simplicity we only mention these topological terms in passing.

This theory is invariant under both parity and charge conjugation. These operators are represented by simple transformations

$$
\begin{aligned}
P &: \ \Sigma \ \to \ \Sigma^{-1}, \\
CP &: \ \Sigma \ \to \ \Sigma^*,
\end{aligned}
\tag{12.34}
$$

where the operation $*$ refers to complex conjugation. The eight meson fields are pseudo-scalars. The neutral pion and the eta meson are both even under charge conjugation.

With massless quarks, the underlying quark-gluon theory has a chiral symmetry under

$$
\begin{aligned}
\psi_L &\to g_L \psi_L, \\
\psi_R &\to g_R \psi_R.
\end{aligned}
\tag{12.35}
$$

Here (g_L, g_L) is in $SU(3) \otimes SU(3)$ and $\psi_{L,R}$ represent the chiral components of the quark fields, with flavor indices understood. This symmetry is expected to be broken spontaneously to a vector SU(3) via a vacuum expectation value for $\overline{\psi}_L \psi_R$. This motivates the sigma model through the identification

$$
\langle 0 | \overline{\psi}_L \psi_R | 0 \rangle \leftrightarrow v\Sigma.
\tag{12.36}
$$

The quantity v, of dimension cubed mass, characterizes the strength of the spontaneous breaking of this symmetry. Thus, the effective field transforms under a chiral symmetry of form

$$
\Sigma \to g_L \Sigma g_R^\dagger.
\tag{12.37}
$$

The Lagrangian density in Eq. (12.33) is the simplest non-trivial expression invariant under this symmetry.

The quark masses break the chiral symmetry explicitly. From the analogy in Eq. (12.36), these are introduced through a 3 by 3 mass matrix M

appearing in a potential term added to the Lagrangian density

$$L = L_0 - v\text{Re Tr}(\Sigma M). \tag{12.38}$$

Here, v is the same dimensionful factor appearing in Eq. (12.36). The chiral symmetry of our starting theory shows the physical equivalence of a given mass matrix M with a rotated matrix $g_R^\dagger M g_L$. Using this freedom, we can put the mass matrix into a standard form [89]. We will assume it is diagonal with increasing eigenvalues

$$M = \begin{pmatrix} m_u & 0 & 0 \\ 0 & m_d & 0 \\ 0 & 0 & m_s \end{pmatrix}, \tag{12.39}$$

representing the up, down, and strange quark masses. Note that this matrix has both singlet and octet parts under flavor symmetry.

$$M = \frac{m_u + m_d + m_s}{3} + \frac{m_u - m_d}{2}\lambda_3 + \frac{m_u + m_d - 2m_s}{2\sqrt{3}}\lambda_8. \tag{12.40}$$

In general, the mass matrix can still be complex. The chiral symmetry allows us to move phases between the masses, but the determinant of M is invariant and physically meaningful. Under charge conjugation, the mass term would only be invariant if $M = M^*$. If $|M|$ is not real, then its phase is the famous CP-violating parameter that we will discuss extensively later. For the moment, however, take all quark masses to be real.

To lowest order, the pseudo-scalar meson masses appear on expanding the mass term quadratically in the meson fields. This generates an effective mass matrix for the eight mesons

$$\mathcal{M}_{\alpha\beta} \propto \text{Re Tr } \lambda_\alpha\lambda_\beta M. \tag{12.41}$$

The isospin breaking up-down mass difference gives this matrix an off-diagonal piece mixing the π_0 and the η

$$\mathcal{M}_{3,8} \propto m_u - m_d. \tag{12.42}$$

The eigenvalues of this matrix give the standard mass relations

$$m_{\pi_0}^2 \propto \frac{2}{3}\Bigg(m_u + m_d + m_s$$

$$- \sqrt{m_u^2 + m_d^2 + m_s^2 - m_u m_d - m_u m_s - m_d m_s} \Bigg),$$

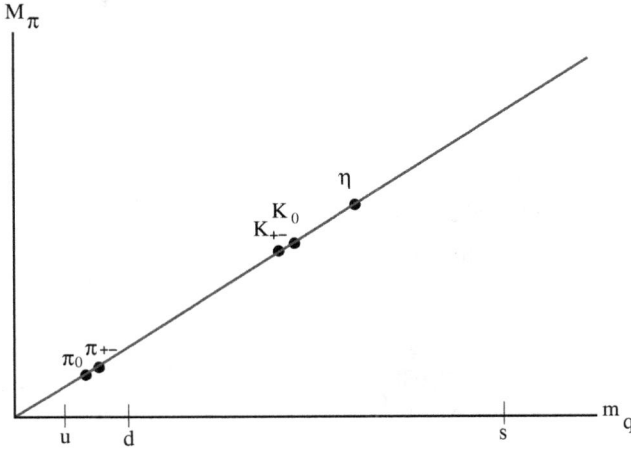

Figure 12.4: The qualitative spectrum for the pseudo-scalar octet three-flavor QCD with unequal quark masses. The neutral pion mass is reduced slightly below the charged states due to isospin breaking and mixing with the eta. The eta itself is a mixture from all three quark species. Effects from the eta prime are ignored here.

$$m_\eta^2 \propto \frac{2}{3}\bigg(m_u + m_d + m_s$$

$$+ \sqrt{m_u^2 + m_d^2 + m_s^2 - m_u m_d - m_u m_s - m_d m_s}\,\bigg),$$

$$m_{\pi_+}^2 = m_{\pi_-}^2 \propto m_u + m_d,$$

$$m_{K_+}^2 = m_{K_-}^2 \propto m_u + m_s,$$

$$m_{K_0}^2 = m_{\overline{K}_0}^2 \propto m_d + m_s. \tag{12.43}$$

Here, we label the mesons with their conventional names. This spectrum is qualitatively sketched in Fig. 12.4.

Redundancies in these relations test the validity of the model. For example, comparing two expressions for the sum of the three quark masses,

$$\frac{2(m_{\pi_+}^2 + m_{K_+}^2 + m_{K_0}^2)}{3(m_\eta^2 + m_{\pi_0}^2)} \sim 1.07, \tag{12.44}$$

suggesting the symmetry should be good to a few percent.

Other combinations of meson masses give estimates for the ratios of the quark masses [90–92]. For one such combination, look at

$$\frac{m_u}{m_d} = \frac{m_{\pi+}^2 + m_{K_+}^2 - m_{K_0}^2}{m_{\pi+}^2 - m_{K_+}^2 + m_{K_0}^2} \sim 0.66. \tag{12.45}$$

This particular combination is polluted by electromagnetic effects; another combination that partially cancels this is

$$\frac{m_u}{m_d} = \frac{2m_{\pi^0}^2 - m_{\pi+}^2 + m_{K_+}^2 - m_{K_0}^2}{m_{\pi+}^2 - m_{K_+}^2 + m_{K_0}^2} \sim 0.55 \tag{12.46}$$

This expression ignores small $m_u m_d / m_s$ corrections in expanding the square root in Eq. (12.43). Shortly, we will comment on a third combination for this ratio. For the strange quark, one can take

$$\frac{2m_s}{m_u + m_d} = \frac{m_{K_+}^2 + m_{K_0}^2 - m_{\pi+}^2}{m_{\pi+}^2} \sim 26. \tag{12.47}$$

Of course, as discussed earlier the quark masses are scale dependent. While their ratios are more stable, we will see later how these ratios also acquire some scale dependence. Nevertheless, from mass differences such as $m_n - m_p \sim 1.3$MeV and $m_{K_0} - m_{K_+} \sim 4.0$MeV, we conclude that the up and down quark masses in these effective models are typically on the order of a few MeV, while the strange quark mass is of order 100 MeV. These are what are known as "current" quark masses, related to chiral symmetries and current algebra. In contrast, since the proton is made of three quarks, some simple quark models consider "constituent" quark masses of a few hundred MeV; these are substantially larger because they include the energy contained in the gluon fields.

While phenomenology, i.e. Eq. (12.46), seems to suggest that the up quark is not massless, there remains a lot of freedom in extracting that ratio from the pseudo-scalar meson masses. From Eq. (12.43), the squared sum of the η and π_0 masses should be proportional to the sum of the three quark masses. Subtracting off the neutral kaon mass should leave just the up quark. Thus motivated, look at

$$\frac{m_u}{m_d} = \frac{3(m_\eta^2 + m_{\pi_0}^2)/2 - 2m_{K_0}^2}{m_{\pi+}^2 - m_{K_+}^2 + m_{K_0}^2} \sim -0.8. \tag{12.48}$$

This strange result is probably a consequence of $SU(3)$ breaking which induces eta and eta-prime mixing, thus lowering the eta mass. But one might worry that depending on what combination of mesons one uses, even the sign of the up quark mass is ambiguous. Attempts to extend the naive quark mass ratio estimates to higher orders in the chiral expansion have shown that there are fundamental ambiguities in the definition of the quark masses [90]. An important message of later sections is that this ambiguity is an inherent property of QCD.

Note that in Eq. (12.43) the squared neutral pion mass can become negative if

$$m_u < \frac{-m_d \bar{m}_s}{m_d + m_s}. \tag{12.49}$$

This situation is inconsistent with experimental results because it will cause a condensation of the pion field and a spontaneous breaking of CP symmetry [93]. This is closely tied to the possibility of a CP-violating term in QCD that we will discuss in later chapters.

Chapter 13

The chiral anomaly

The picture of pions as approximate Goldstone bosons is, of course, completely standard. It is also common lore that the anomaly prevents the η' from being a Goldstone boson and leaves it with a mass on the order of Λ_{QCD}, even in the massless quark limit. The issue is that the effective potential V considered as a function of the fields in Eq. (12.23) must not be symmetric under an anomalous rotation between η' and σ, that is,

$$\begin{aligned} \sigma &\to \sigma \cos(\phi) + \eta' \sin(\phi), \\ \eta' &\to -\sigma \sin(\phi) + \eta' \cos(\phi). \end{aligned} \tag{13.1}$$

In the next section, we discuss how this symmetry is lost, as well as its connection to the zero modes of the Dirac operator.

If we consider the effective potential as a function of the fields σ and η', it should have a minimum at $\sigma \sim v$ and $\eta' \sim 0$. Expanding about that point, we expect a qualitative form

$$V(\sigma, \eta') \sim m_\sigma^2 (\sigma - v)^2 + m_{\eta'}^2 \eta'^2 + O((\sigma - v)^3, \eta'^4). \tag{13.2}$$

We expect both m_σ and $m_{\eta'}$ to remain on the order of Λ_{QCD}, even in the chiral limit. And, at least with an even number of flavors as we are currently considering, there should be a second minimum with $\sigma \sim -v$.

At this point, one can ask whether we know anything else about the effective potential in the (σ, η') plane. We will shortly see that indeed we do, and the potential has a total of N_f equivalent minima in the chiral limit. But first we review how the above minima arise in quark language.

13.1.   What broke the symmetry?

The classical QCD Lagrangian has a symmetry under a rotation of the underlying quark fields

$$\psi \rightarrow e^{i\phi\gamma_5/2}\psi,$$
$$\overline{\psi} \rightarrow \overline{\psi}e^{i\phi\gamma_5/2}. \tag{13.3}$$

This corresponds directly to the transformation of the composite fields given in Eq. (13.1). This symmetry is "anomalous" in that any regulator must break it in a way that leaves a remnant surviving as the regulator is removed [94–96]. With the lattice, this concerns the continuum limit.

The specifics of how the anomaly works depend on the details of the regulator. Here we will follow Fujikawa [97] and consider the fermionic measure in the path integral. If we make the above rotation on the field ψ, the measure changes by the determinant of the rotation matrix

$$d\psi \rightarrow |e^{-i\phi\gamma_5/2}|d\psi = e^{-i\phi\text{Tr}\gamma_5/2}d\psi. \tag{13.4}$$

Here is where the subtlety of the regulator comes in. Naively, γ_5 is a simple four by four traceless matrix. If it is indeed traceless, then the measure would be invariant. However, in the regulated theory this is not the case. This is intimately tied with the index theorem for the Dirac operator in topologically non-trivial gauge fields.

A typical Dirac action takes the form $\overline{\psi}(D+m)\psi$ with the kinetic term D a function of the gauge fields. In the naive continuum theory, D is anti-Hermitian, $D^\dagger = -D$, and anti-commutes with γ_5, i.e. $[D, \gamma_5]_+ = 0$. What complicates the issue with fermions is the index theorem discussed earlier and reviewed in Ref. [98]. If a background gauge field has winding ν, then there must be at least ν exact zero eigenvalues of D. Furthermore, on the space spanned by the corresponding eigenvectors, γ_5 can be simultaneously diagonalized with D. The net winding number equals the number of positive eigenvalues of γ_5 minus the number of negative eigenvalues. In this subspace the trace of γ_5 does not vanish, but equals ν.

What about the higher eigenvalues of D? We discussed these earlier when we formulated the index theorem. Because $[D, \gamma_5]_+ = 0$, non-vanishing eigenvalues appear in opposite sign pairs; i.e. if $D|\psi\rangle = \lambda|\psi\rangle$ then $D\gamma_5|\psi\rangle = -\lambda\gamma_5|\psi\rangle$. For an anti-Hermitian D, these modes are orthogonal with $\langle\psi|\gamma_5|\psi\rangle = 0$. As a consequence, γ_5 is traceless on the subspace spanned by each pair of eigenvectors.

So, as the zero modes appear with changing topology, what happens to their chiral partners? In a regulated theory they are, in some sense, "above the cutoff." In a simple continuum discussion, they have been "lost at infinity." With a lattice regulator, there is no "infinity." Something more subtle must happen. With the overlap [99, 100] or Wilson [13] fermions, discussed in more detail later, one gives up the anti-Hermitian character of D. Most eigenvalues still occur in conjugate pairs and do not contribute to the trace of γ_5. However, in addition to small real eigenvalues representing the zero modes, there are additional modes where the eigenvalues are also real, but large. With Wilson fermions, these appear as massive doubler states. With the overlap, the eigenvalues are constrained to lie on a circle. In this case, for every exact zero mode there is another mode with the opposite chirality lying on the opposite side of the circle. These modes are effectively massive and break chiral symmetry. The necessary involvement of both small and large eigenvalues warns of the implicit danger in attempts to separate infrared from ultraviolet effects. Where the anomaly is concerned, going to short distances is not sufficient for ignoring non-perturbative effects related to topology.

So with the regulator in place, the trace of γ_5 does not vanish on gauge configurations of non-trivial topology. The change of variables indicated in Eq. (13.4) introduces into the path integral a modification of the weighting by a factor

$$e^{-i\phi \mathrm{Tr}\gamma_5} = e^{-i\phi N_f \nu}. \tag{13.5}$$

Here we have applied the rotation to all flavors equally, thus the presence of the factor of N_f in the exponent. The conclusion is that gauge configurations that have non-trivial topology receive a complex weight after the anomalous rotation. By treating the N_f flavors equivalently, here we divide the conventionally defined CP violation angle Θ — to be discussed later — equally among the flavors, *i.e.* $\phi = \Theta/N_f$ effectively. Although not the topic of discussion here, note that this factor introduces a sign problem for Monte Carlo simulations.

13.2. A discrete chiral symmetry

We now return to the effective Lagrangian language of before. For the massless theory, the symmetry under $\sigma \leftrightarrow -\sigma$ indicates that we expect at least two minima for the effective potential considered in the σ, η' plane, as suggested in Fig. 13.1. Do we know anything about the potential elsewhere

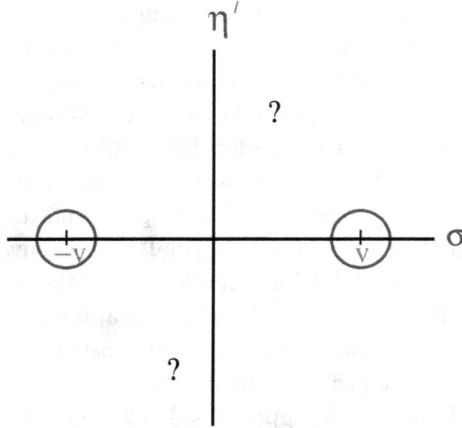

Figure 13.1:　We have two minima in the σ, η' plane located at $\sigma = \pm v$ and $\eta' = 0$. The circles represent that the fields will fluctuate in a small region about these minima. Can we find any other minima?

in this plane? The answer is yes, indeed there are actually N_f equivalent minima.

It is convenient to separate the left and right hand parts of the fermion field, so that

$$\psi_{L,R} = \frac{1}{2}(1 \pm \gamma_5)\psi,$$

$$\overline{\psi}_{L,R} = \overline{\psi}\frac{1}{2}(1 \mp \gamma_5). \tag{13.6}$$

The mass term is thus

$$m\overline{\psi}\psi = m(\overline{\psi}_L\psi_R + \overline{\psi}_R\psi_L) \tag{13.7}$$

and mixes the left and right components.

Using this notation, due to the anomaly, the singlet rotation

$$\psi_L \rightarrow e^{i\phi}\psi_L \tag{13.8}$$

is not a valid symmetry of the theory for generic values of the angle ϕ. On the other hand, flavored chiral symmetries should survive, and in particular,

$$\psi_L \rightarrow g_L\psi_L = e^{i\phi_\alpha \lambda_\alpha}\psi_L \tag{13.9}$$

is expected to be a valid symmetry for any set of angles ϕ_α. The point of this section is that, for special special discrete values of the angles, the

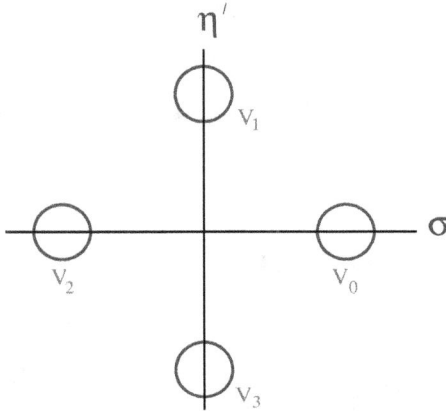

Figure 13.2: For four massless flavors we have four equivalent minima in the σ, η' plane. With N_f flavors, this generalizes to N_f distinct minima.

rotations in Eqs. (13.8) and (13.9) can coincide. At such values, the singlet rotation becomes a valid symmetry. In particular, note that

$$g = e^{2\pi i \phi / N_f} \in Z_{N_f} \subset SU(N_f). \tag{13.10}$$

Thus, a valid discrete symmetry involving only σ and η' is

$$\begin{aligned} \sigma &\to \quad \sigma \cos(2\pi/N_f) + \eta' \sin(2\pi/N_f), \\ \eta' &\to -\sigma \sin(2\pi/N_f) + \eta' \cos(2\pi/N_f). \end{aligned} \tag{13.11}$$

The potential $V(\sigma, \eta')$ has a Z_{N_f} symmetry manifested in N_f equivalent minima in the (σ, η') plane. This structure is sketched in Fig. 13.2 for the four-flavor case.

This discrete flavor singlet symmetry arises from the trivial fact that Z_N is a subgroup of both $SU(N)$ and $U(1)$. At the quark level the symmetry is easily understood, since the quark measure receives an additional phase proportional to the winding number from every flavor. With N_F flavors, these multiply together, making

$$\psi_L \to e^{2\pi i / N_f} \psi_L \tag{13.12}$$

a valid symmetry even though rotations by smaller angles are not.

The role of the Z_N center of $SU(N)$ is illustrated graphically in Fig. 13.3, taken from Ref. [101]. Here we plot the real and the imaginary parts of the traces of 10,000 $SU(3)$ matrices drawn randomly with the invariant group measure. The region of support only touches the $U(1)$ circle at the elements

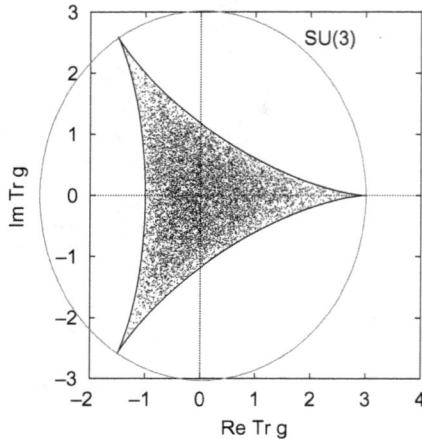

Figure 13.3:   The real and imaginary parts for the traces of 10,000 randomly-chosen $SU(3)$ matrices. All points lie within the boundary, representing matrices of the form $\exp(i\phi\lambda_8)$. The tips of the three points represent the center of the group. The outer curve represents the boundary that would be found if the group was the full $U(1)$. (Taken from Ref. [101]).

of the center. All elements lie on or within the curve mapped out by elements of the form $\exp(i\phi\lambda_8)$. The first order transition at $\Theta = \pi$ can be thought of as a jump in the typical sigma field from one of the points in the figure to another. Figure 13.4 is a similar plot for the group $SU(4)$.

13.3.   The 't Hooft vertex

The consequences of non-trivial gauge topology and the connections to the anomaly are often described in terms of an effective multi-fermion interaction referred to as the "'t Hooft vertex." To understand the 't Hooft interaction in path integral language, we begin with a reminder of the underlying strategy of lattice simulations. Consider the generic path integral, or "partition function," for quarks and gluons:

$$Z = \int (dA)(d\psi \ d\overline{\psi}) \exp\left(-S_g(A) - \overline{\psi}D(A)\psi\right). \qquad (13.13)$$

Here A denotes the gauge fields and $\overline{\psi}, \psi$ the quark fields. The pure gauge part of the action is $S_g(A)$ and the matrix describing the fermion part of the action is $D(A)$. Since direct numerical evaluation of the fermionic integrals appears to be impractical, the Grassmann integrals are conventionally

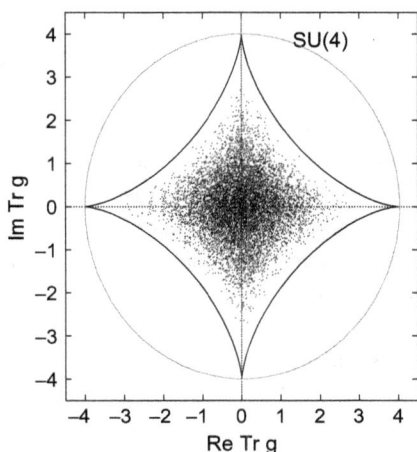

Figure 13.4: The generalization of Fig. 13.3 to $SU(4)$. The real and imaginary parts for the traces of 10,000 randomly-chosen $SU(4)$ matrices. (Taken from Ref. [101]).

evaluated analytically, reducing the partition function to

$$Z = \int (dA)\, e^{-S_g(A)}\, |D(A)|, \qquad (13.14)$$

where $|D(A)|$ denotes the determinant of the Dirac matrix evaluated in the given gauge field. Thus motivated, the basic lattice approach is to generate a set of random gauge configurations weighted by $\exp(-S_g(A))\, |D(A)|$. Given an ensemble of such configurations, one then estimates physical observables by averages over this ensemble.

This procedure seems innocent enough, but it can run into trouble when one has massless fermions and corresponding zero modes associated with topology. To see the issue, write the determinant as a product of the eigenvalues λ_i of the matrix D. In general, D may not be a normal matrix; so, one should pick either left or right eigenvectors at one's discretion. This is a technical detail that will not play any further role here. In order to control infrared issues with massless quarks, introduce a small explicit mass m and reduce the path integral to

$$Z = \int (dA)\, e^{-S_g(A)} \prod (\lambda_i + m). \qquad (13.15)$$

Now, suppose we have a configuration where one of the eigenvalues of $D(A)$ vanishes, *i.e.* assume that some $\lambda_i = 0$. This, of course, is what happens with non-trivial topology present. As we take the mass to zero, any configurations involving such an eigenvalue will drop out of the ensemble. At first, one might suspect this would be a set of measure zero in the space of all possible gauge fields. However, as discussed above, the index theorem ties gauge field topology to such zero modes. In general, these modes are robust under small deformations of the fields. Under the traditional lattice strategy, the corresponding configurations would then have zero weight in the massless limit. The naive conclusion is that such configurations are irrelevant to physics in the chiral limit.

It was this reasoning that 't Hooft showed to be incorrect. Indeed, he demonstrated that it is natural for some observables to have $1/m$ factors when zero modes are present. These can cancel the terms linear in m from the determinant, leaving a finite contribution.

As a simple example, consider the quark condensate in one-flavor QCD

$$\langle \overline{\psi}\psi \rangle = \frac{1}{VZ} \int (dA) \; e^{-S_g} \; |D| \; \text{Tr}D^{-1}. \tag{13.16}$$

Here V represents the system volume, inserted to give an intensive quantity. Expressing the fermionic factors in terms of the eigenvalues of D reduces this to

$$\langle \overline{\psi}\psi \rangle = \frac{1}{VZ} \int (dA) \; e^{-S_g} \left(\prod_i (\lambda_i + m) \right) \sum_j \frac{1}{\lambda_j + m}. \tag{13.17}$$

Now, if there is a mode with $\lambda_i = 0$, the factor of m is canceled by a $1/m$ piece in the trace of D^{-1}. Configurations containing a zero mode give a constant contribution to the condensate, and this contribution survives in the massless limit. Note that this effect is unrelated to spontaneous breaking of chiral symmetry and appears even with finite volume.

This contribution to the condensate is special to the one-flavor theory. Because of the anomaly, this quark condensate is not an order parameter for any symmetry. With more fermion species there will be additional factors of m from the determinant. Then, the effect of the 't Hooft vertex is of higher order in the fermion fields and does not appear directly in the condensate. For two or more flavors, the standard Banks-Casher picture [102] of an eigenvalue accumulation leading to the spontaneous breaking of chiral symmetry should apply.

The conventional discussion of the 't Hooft vertex starts by inserting fermionic sources into the path integral

$$Z(\eta, \overline{\eta}) = \int (dA) \; (d\psi) \; (d\overline{\psi}) \; e^{-S_g - \overline{\psi}(D+m)\psi + \overline{\psi}\eta + \overline{\eta}\psi}. \tag{13.18}$$

Differentiation, in the Grasmannian sense, with respect to these sources will generate the expectation for an arbitrary product of fermionic operators. Integrating out the fermions reduces this to

$$Z = \int (dA) \, e^{-S_g + \overline{\eta}(D + m)^{-1}\eta} \prod (\lambda_i + m). \tag{13.19}$$

Consider a zero mode ψ_0 satisfying $D\psi_0 = 0$. In general, there is also a left zero mode satisfying $\overline{\psi}_0 D = 0$. If the sources have an overlap with the mode, that is $(\overline{\eta}|\psi_0) \neq 0$, then a factor of $1/m$ in the source term can cancel the m from the determinant. Although non-trivial topological configurations do not contribute to Z, their effects can survive in correlation functions. For the one-flavor theory the effective interaction is bilinear in the fermion sources and is proportional to

$$(\overline{\eta}|\psi_0)(\overline{\psi}_0|\eta). \tag{13.20}$$

As discussed earlier, the index theorem tells us that, in general, the zero mode is chiral; it appears in either $\overline{\eta}_L \eta_R$ or $\overline{\eta}_R \eta_L$, depending on the sign of the gauge field winding.

With $N_f \geq 2$ flavors, the cancellation of the mass factors in the determinant requires source factors from each flavor. This combination is the 't Hooft vertex. It is an effective $2N_f$ fermion operator. In the process, every flavor flips its spin, as sketched in Fig. 13.5. Indeed, this is the chiral anomaly; left and right helicities are not separately conserved.

Because of Pauli statistics, the multi-flavor vertex can be written in the form of a determinant. This clarifies how the vertex preserves flavored chiral symmetries. With two flavors, u and d, Eq. (13.20) generalizes to

$$\begin{vmatrix} (\overline{u}|\psi_0)(\overline{\psi}_0|u) & (\overline{u} \, \psi_0)(\overline{\psi}_0|d) \\ (\overline{d}|\psi_0)(\overline{\psi}_0|u) & (\overline{d} \, \psi_0)(\overline{\psi}_0|d) \end{vmatrix}. \tag{13.21}$$

The general form of the vertex involves factors from each flavor of quark. This provides an alternative way to see the Z_{N_f} chiral symmetry discussed earlier in this chapter.

Note that the effect of the vertex is non-local. In general, the zero mode ψ_0 is spread out over the finite region of the "instanton", i.e. the size parameter ρ from the explicit solution given earlier. This means there is an inherent position-space uncertainty concerning where the fermions are interacting. A particular consequence is that fermion conservation is only a global symmetry. In Minkowski space language, this non-locality can be thought of in terms of states sliding in and out of the Dirac sea at different locations.

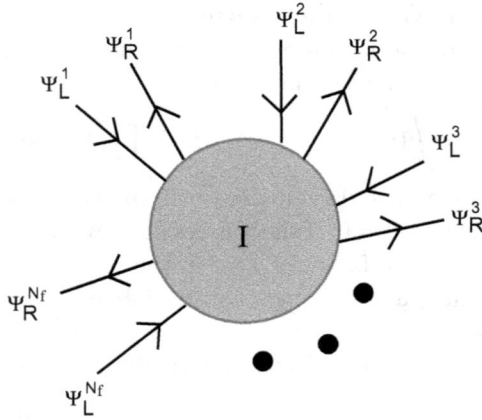

Figure 13.5: The 't Hooft vertex for N_f flavors is a $2N_f$ effective fermion operator that flips the spin of every flavor.

13.4. Fermions in higher representations

When the quarks are massless, the classical field theory corresponding to the strong interactions has a $U(1)$ axial symmetry under the transformation

$$\psi \to e^{i\theta\gamma_5}\psi \qquad \overline{\psi} \to \overline{\psi}e^{i\theta\gamma_5}. \qquad (13.22)$$

It is the 't Hooft vertex that explicitly demonstrates how this symmetry does not survive quantization. This multi-fermion vertex contains a fermion bilinear for every zero mode from the index theorem. Because it is often a high order fermion operator, interesting discrete chiral symmetries can remain. This is particularly true when the fermions are higher representations of the gauge group. In this section, we explore how quarks in non-fundamental representations can display unexpected discrete chiral subgroups.

While these considerations do not apply to the usual theory of the strong interactions where the quarks are in the fundamental representation, there are several reasons to study them anyway. At higher energies, perhaps as being probed at the Large Hadron Collider, one might well discover new strong interactions that play a substantial role in the spontaneous breaking of the electroweak theory. Also, many grand unified theories involve fermions in non-fundamental representations. As one example, we will see that massless fermions in the 10 representation of $SU(5)$ possess a Z_3 discrete chiral symmetry. Similarly, the left handed 16 covering representation of $SO(10)$ gives a chiral gauge theory with a surviving discrete Z_2 chiral symmetry. Understanding these symmetries may eventually play a role in a discretization of chiral gauge theories on the lattice.

To start, we generalize the index theorem relating gauge field topology to zero modes of the Dirac operator. In particular, fermions in higher representations can involve multiple zero modes for a given winding. Being generic, consider representation X of a gauge group G. Denote by N_X the number of zero modes that are required per unit of winding number in the gauge fields. That is, suppose the index theorem generalizes to

$$n_r - n_l = N_X \nu \qquad (13.23)$$

where n_r and n_l are the number of right and left handed zero modes, respectively, and ν is the winding number of the associated gauge field. The basic 't Hooft vertex receives contributions from each zero mode, resulting in an effective operator which is a product of $2N_X$ fermion fields. Schematically, the vertex is modified along the lines $\overline{\psi}_L \psi_R \longrightarrow (\overline{\psi}_L \psi_R)^{N_X}$. While this form still breaks the $U(1)$ axial symmetry, it is invariant under $\psi_R \to e^{2\pi i/N_X} \psi_R$. In other words, there is a Z_{N_X} discrete axial symmetry.

There are a variety of convenient tools for determining N_X. Consider building up representations from lower ones. Take two representations X_1 and X_2 and form the direct product representation $X_1 \otimes X_2$. Let the matrix dimensions for X_1 and X_2 be D_1 and D_2, respectively. Then, for the product representation we have

$$N_{X_1 \otimes X_2} = N_{X_1} D_{X_2} + N_{X_2} D_{X_1}. \qquad (13.24)$$

To see this, start with X_1 and X_2 representing two independent groups G_1 and G_2. With G_1 having winding, there will be a zero mode for each of the dimensions of the matrix index associated with X_2. Similarly, there will be multiple modes for winding in G_2. These modes are robust and all should remain if we now constrain the groups to be the same.

As a first example, denote the fundamental representation of $SU(N)$ as F and the adjoint representation as A. Then using $\overline{F} \otimes F = A \oplus 1$ in the above gives $N_A = 2N_F$, as noted some time ago [103]. With $SU(3)$, fermions in the adjoint representation will have six-fold degenerate zero modes.

For another example, consider $SU(2)$ and build up towards arbitrary spin $s \in \{0, \frac{1}{2}, 1, \frac{3}{2}, \ldots\}$. Recursing the above relation gives the result for arbitrary spin

$$N_s = s(2s+1)(2s+2)/3. \tag{13.25}$$

Another technique for finding N_X in more complicated groups begins by rotating all topological structure into an $SU(2)$ subgroup and then counting the corresponding $SU(2)$ representations making up the larger representation of the whole group. An example to illustrate this procedure is the anti-symmetric two indexed representation of $SU(N)$. This representation has been extensively used in Refs. [104–107] for an alternative approach to the large gauge group limit. The basic $N(N-1)/2$ fermion fields take the form

$$\psi_{ab} = -\psi_{ba}, \qquad a, b \in 1, 2, \ldots N. \tag{13.26}$$

Consider rotating all topology into the $SU(2)$ subgroup involving the first two indices, i.e. 1 and 2. Because of the anti-symmetrization, the field ψ_{12} is a singlet in this subgroup. The field pairs $(\psi_{1,j}, \psi_{2,j})$ form a doublet for each $j \geq 3$. Finally, the $(N-2)(N-3)/2$ remaining fields do not transform under this subgroup and are singlets. Overall, we have $N-2$ doublets under the $SU(2)$ subgroup, each of which gives one zero mode per winding number. We conclude that the 't Hooft vertex leaves behind a Z_{N-2} discrete chiral symmetry. Specializing to the 10 representation of $SU(5)$, this is the Z_3 mentioned earlier.

A further example is the group $SO(10)$ with fermions in the 16 dimensional covering group. This forms the basis of a rather interesting grand unified theory, where one generation of fermions is placed into a single left handed 16 multiplet [108]. This representation includes two quark species interacting with the $SU(3)$ subgroup of the strong interactions. Rotating a topological excitation into this subgroup, we see that the effective vertex will be a four-fermion operator and preserve a Z_2 discrete chiral symmetry.

It is unclear whether these discrete symmetries are expected to be spontaneously broken. Since they are discrete, such breaking is not associated with Goldstone bosons. But the quark condensate does provide an order parameter; so, when $N_X > 1$, any such breaking would be conceptually meaningful. This could be checked in numerical simulations.

Further study

- Unlike with the η meson, the neutral vector meson ϕ shows little mixing with light quarks [3, 109, 110]. In this case, $\bar{s}\gamma_\mu s$ is a better interpolating operator than $\bar{\psi}\lambda_8\psi$. Why do the vector mesons behave so differently from the pseudo-scalars?

Chapter 14

Massive quarks and the Theta parameter

As discussed earlier and illustrated in Fig. 13.2, a quark mass term $m\overline{\psi}\psi \sim m\sigma$ is represented by a "tilting" of an effective potential. This selects one of the multiple minima in the σ, η' plane as the true vacuum. When the quark masses are small in comparison to the basic scale of QCD, the other minima will persist as extrema, although some of them may become unstable under small fluctuations due to the flatness of flavor-non-singlet directions. Counting the minima sequentially with the true vacuum having $n = 0$, each is associated with excitations in the pseudo-Goldstone directions that have an effective mass of $m_\pi^2 \sim m \cos(2\pi n/N_f)$. Note that when N_f exceeds four, there will be more than one meta-stable state. However, in the usual case of considering two or three quarks as light, only one minimum remains locally stable.

14.1. Twisted tilting

Conventionally, the mass tilts the potential downward in the positive σ direction. However, it is an interesting exercise to consider tilts in other directions in the σ, η' plane. This is accomplished with an anomalous rotation on the mass term

$$m\overline{\psi}\psi \to m\cos(\phi)\overline{\psi}\psi + im\sin(\phi)\overline{\psi}\gamma_5\psi$$

$$\sim m\cos(\phi)\sigma + m\sin(\phi)\eta'. \tag{14.1}$$

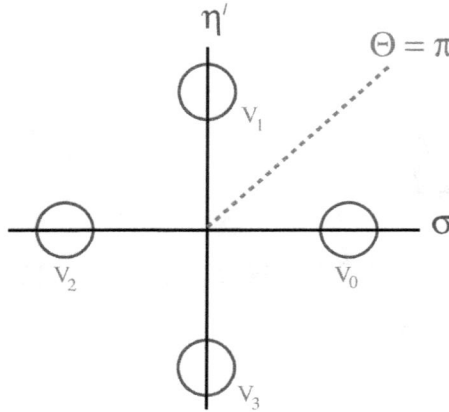

Figure 14.1: With massive quarks and a twisting angle of $\phi = \pi/N_f$, two of the minima in the σ, η' plane become degenerate. This corresponds to a first order transition at $\Theta = \pi$.

Were it not for the anomaly, this would just be a redefinition of fields. However, the same effect that gives the η' its mass indicates that this new form for the mass term gives an inequivalent theory. As $i\bar{\psi}\gamma_5\psi$ is odd under CP, this theory is explicitly CP-violating.

The conventional notation for this effect involves the angle $\Theta = N_f\phi$. Then the Z_{N_f} symmetry amounts to a 2π periodicity in Θ. As Fig. 14.1 indicates two degenerate minima exist at special values of the twisting angle ϕ. This occurs, for example, at $\phi = \pi/N_f$ or $\Theta = \pi$. As the twisting increases through this point, there will be a first order transition as the true vacuum jumps from the vicinity of one minimum to the next.

Because of the Z_{N_f} symmetry of the massless theory, all the N_f separate minima are physically equivalent. This means that if we apply our mass term in the direction of any of them, we obtain the same theory. In particular, for four flavors the usual mass term $m\bar{\psi}\psi$ is equivalent to using the alternative mass term $im\bar{\psi}\gamma_5\psi$. This result, however, is true if and only if N_f is a multiple of four.

14.2. Odd N_f

One interesting consequence of this way of looking at the Theta parameter concerns QCD with an odd number of flavors. The group $SU(N_f)$ with odd N_f does not include the element -1. In particular, the Z_{N_f} structure

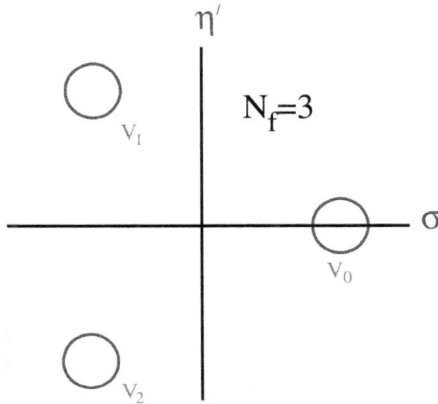

Figure 14.2: For odd N_f, such as the three-flavor case sketched here, QCD is not symmetric under a change in sign of the quark mass. Negative mass corresponds to taking $\Theta = \pi$.

is not symmetric under reflections about the η' axis. Figure 14.2 sketches the situation for $SU(3)$. One immediate conclusion is that positive and negative mass are not equivalent. Indeed, a negative mass with three degenerate flavors corresponds to the $\Theta = \pi$ case, where a spontaneous breaking of CP is expected. In this case there is no symmetry under flipping the sign of $\sigma \sim \bar{\psi}\psi$. The simple picture sketched in Fig. 12.1 no longer applies.

At $\Theta = \pi$, the theory lies on top of a first order phase transition line. A simple order parameter for this transition is the expectation value for the η' field. As this field is odd under CP symmetry, this shows that negative mass QCD with an odd number of flavors spontaneously breaks CP.[1] This does not contradict the Vafa-Witten theorem [111] because in this regime the fermion determinant is not positive definite.

Note that the asymmetry in the sign of the quark mass is not easily seen in perturbation theory. Any quark loop in a perturbative diagram can have the sign of the quark mass flipped by a γ_5 transformation. It is only through the subtleties of regulating the divergent triangle diagram [94–96] that the sign of the mass enters.

A remarkable conclusion of these observations is that two physically distinct theories can have identical perturbative expansions. For example,

---

[1] Dashen's original paper [25] speculates that this might be related to the parity breaking seen in nature. This presumably requires a new "beyond the Standard Model" interaction rather than QCD.

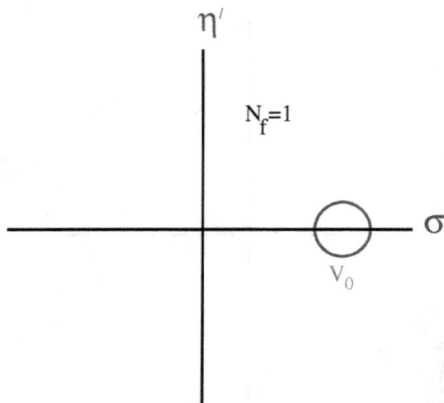

Figure 14.3:   The effective potential for one-flavor QCD with small quark mass has a unique minimum in the σ, η' plane. The minimum is shifted from zero due to the effect of the 't Hooft vertex.

with three flavors the negative mass theory has spontaneous CP violation, while the positive mass theory does not. Yet both cases have exactly the same perturbation theory. This dramatically demonstrates what we already knew: non-perturbative effects are essential to understanding QCD.

A special case of an odd number of flavors is one-flavor QCD [112]. In this case the anomaly removes all traces of chiral symmetry and there is a unique minimum in the σ, η' plane, as sketched in Fig. 14.3. This minimum does not occur at the origin, being shifted to $\langle \bar\psi\psi \rangle > 0$ by the 't Hooft vertex, which for one flavor is just an additive mass shift [113]. Unlike the case with more flavors, the resulting expectation value for σ is not from a spontaneous symmetry breaking; indeed, there is no chiral symmetry to break. Any regulator that preserves a remnant of chiral symmetry must inevitably fail [114]. Note also that for one-flavor QCD, there is no longer the necessity of a first order phase transition at $\Theta = \pi$. Indeed, the region around vanishing mass is not expected to show any singularity. However, such a transition is expected if the mass is sufficiently negative [112]. This can be seen by taking a second flavor to large mass in the two-flavor theory, discussed in the next chapter.

An unusual feature of one-flavor QCD is that the renormalization of the quark mass is not multiplicative when non-perturbative effects are taken into account. The additive mass shift is generally scheme dependent since the details of the instanton effects depend on scale. This is the basic reason that a massless up quark is not a possible solution to the strong CP problem

[115]. Later we will discuss this in more detail in the context of the two-flavor theory with non-degenerate masses.

Because of this shift in the mass, the conventional parameters Θ and m are singular coordinates for the one-flavor theory. A cleaner set of variables is to consider directly the coefficients of the two possible mass terms $\bar{\psi}\psi$ and $i\bar{\psi}\gamma_5\psi$ in the Lagrangian. The issue is a possible ambiguity in the quark mass and is closely related to the necessary existence of rough gauge field configurations. When the gauge fields are not smooth, the presence of a zero mode can depend on the detailed fermion operator in use. This applies even to the formally elegant overlap operator that we will discuss later. Smoothness conditions imposed on the gauge fields to remove this ambiguity are known to conflict with fundamental principles, such as reflection positivity [116].

The Z_{N_f} symmetry discussed here is a property of the fermion determinant and is independent of the gauge field dynamics. In Monte Carlo simulation language, this symmetry appears configuration by configuration. With N_f flavors, we always have $|D| = |e^{2\pi i/N_f}D|$ for any gauge field and any values for the individual quark masses. Depending non-trivially on N_f, this discrete chiral symmetry is inherently discontinuous in N_f. This non-continuity lies at the heart of the issues with the rooted staggered quark approximation. We will return to this topic in Chapter 17.

The arguments in this chapter have relied on chiral symmetry and the presence of dynamical fermions. This leaves open the question of whether a first order transition at $\Theta = \pi$ persists in the pure gauge theory, i.e. $N_f = 0$. This can be inferred by increasing quark mass in the N_f flavor case towards infinity. Above we have considered the expectation of $i\bar{\psi}\gamma_5\psi$ as an order parameter. At intermediate masses this operator will mix with the pseudo-scalar operator $F\tilde{F}$, giving the latter an expectation value as well. This provides a natural order parameter for the quarkless theory.

The only way to lose the jump at π is to have an additional transition in the mass parameter that reduces the degenerate ground states at $\Theta = \pi$ to a unique vacuum. Without some motivation, this seems unlikely. An old argument of 't Hooft for a first order transition at $\Theta = \pi$ in the pure gauge theory appears in Ref. [117], recently elaborated on formally in Ref. [118].

Further study

- What would the analog of Fig. 13.3 or 13.4 look like for the group $SU(2)$? Do we still expect a first order transition at $\Theta = \pi$?

Chapter 15

Quark masses in two-flavor QCD

In the previous chapter we concentrated on degenerate quarks. In general, each species introduces another complex mass parameter. Using flavored chiral rotations, we can move the phases of the masses around arbitrarily, leaving only one overall phase, the Theta (Θ) parameter. Once the overall scale has been set, QCD depends on $N_f + 1$ parameters.

Here we explore the rich phase diagram of two-flavor QCD as a function of the most general quark masses, including the Θ parameter. This chapter closely follows the discussion in Ref. [119]. The theory involves three independent parameters. One is CP-violating; the strong experimental limit on this is the strong CP problem. Here we will characterize the parameters by distinguishing their transformations under various symmetries. As we define them, the resulting variables are each multiplicatively renormalized. However, non-perturbative effects are not universal, leaving individual quark mass ratios with a renormalization scheme dependence. This exposes ambiguities in matching lattice results with perturbative schemes and the tautology involved in approaches that attack the strong CP problem via a vanishing mass for the lightest quark.

Before turning on the masses, we re-emphasize the qualitative properties expected in massless two-flavor QCD. Of course, being an interacting quantum field theory, nothing has been proven rigorously. While the classical theory is conformally invariant, as discussed earlier, confinement and dimensional transmutation generate a non-trivial mass scale. The quantum theory should contain massive stable nucleons. On the other hand, spontaneous chiral symmetry breaking should give rise to three massless pions as Goldstone bosons. Bound states of glue will in general acquire a finite width due to decays into pions. In addition, the two-flavor analog of the eta-prime meson should acquire its mass from the anomaly.

In this theory, the eta-prime and neutral pion involve distinct combinations of quark-antiquark pairs. In the simple quark model, the neutral pseudo-scalars involve the combinations

$$\pi_0 \sim \bar{u}\gamma_5 u - \bar{d}\gamma_5 d,$$

$$\eta' \sim \bar{u}\gamma_5 u + \bar{d}\gamma_5 d + \text{glue}. \tag{15.1}$$

Here, we include a gluonic contribution from mixing between the η' and glueball states. When the quarks are degenerate, isospin forbids such mixing for the pion.

Projecting out helicity states for the quarks $q_{L,R} = (1 \pm \gamma_5)q/2$, the pseudo-scalars are combinations of left with right handed fermions, *i.e.* $\bar{q}_L q_R - \bar{q}_R q_L$. Thus, as shown schematically in Fig. 15.1, meson exchange will contribute to a spin flip process in a hypothetical quark scattering experiment. The four-point function $\langle \bar{u}_R u_L \bar{d}_R d_L \rangle$ does not vanish. Scalar meson exchange will also contribute to this process, but this is not important for the qualitative argument below. Of course, we must assume that some sort of gauge fixing has been done to eliminate a trivial vanishing of this function from an integral over gauges. We also consider this four-point function at a scale before confinement sets in.

It is important that the π_0 and η' are not degenerate. This is due to the anomaly and the fact that the η' is not a Goldstone boson. As discussed earlier, the π_0-η' mass difference can be ascribed to topological structures in the gauge field. Because the mesons are not degenerate, their contributions to the above diagram cannot cancel. The conclusion of this simple argument is that helicity-flip quark-quark scattering is not suppressed in the chiral limit.

Now consider turning on a small down quark mass while leaving the up quark massless. Formally, such a mass allows one to connect the ingoing and outgoing down-quark lines in Fig. 15.1, thereby inducing a mixing between the left and right handed up quark. Such a process is sketched in Fig. 15.2.

So the presence of a non-zero d-quark mass will induce an effective mass for the u quark, even if the latter initially vanishes. As a consequence, non-perturbative effects renormalize m_u/m_d. If this ratio is zero at some scale, it cannot remain so for all scales. Only in the isospin limit are quark mass ratios renormalization group invariant. As lattice simulations

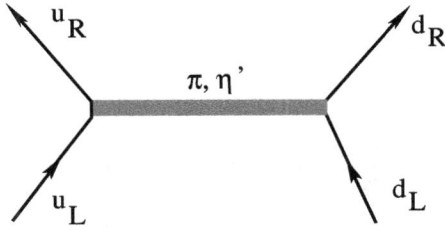

Figure 15.1: Both pion and eta-prime exchange contribute to spin flip scattering between up and down quarks. (Figure from Ref. [119]).

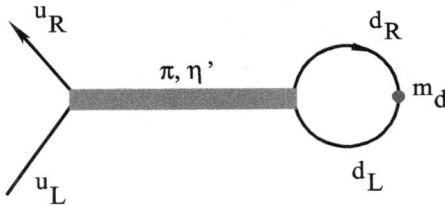

Figure 15.2: Through physical meson exchange, a down quark mass can induce an effective mass for the up quark.

include all perturbative and non-perturbative effects, this phenomenon is automatically included in such an approach.

This cross talk between the masses of different quark species is a relatively straightforward consequence of the chiral anomaly and has been discussed several times in the past, usually in the context of gauge field topology and the index theorem [115, 120–122]. This mixing does not occur in perturbation theory. Feynman diagrams suppress spin-flip processes as the quark masses go to zero. The above argument shows that this lore need not apply when anomalous processes come into play. In particular, mass renormalization is not flavor blind and the concept of mass independent regularization is problematic. Since the quark masses influence each other, there are inherent ambiguities defining $m_u = 0$. This has consequences for the strong CP problem, discussed further below. Furthermore, a traditional perturbative regulator such as $\overline{MS}$ is not complete when $m_u \neq m_d$. Because of this, the practice of matching lattice calculations to $\overline{MS}$ is ill-defined.

Despite the deceptive simplicity of the above argument, it is important to realize that it directly contradicts perturbative lore. At this point, a variety of questions arise.

Can one avoid these questions by working directly with bare quark masses? This runs into problems since the bare masses all vanish under renormalization. Earlier we discussed the renormalization group equation for a quark mass

$$a\frac{dm_i}{da} = \gamma(g)m_i = \gamma_0 g^2 + O(g^4).\qquad(15.2)$$

As asymptotic freedom drives the bare coupling to zero, the bare masses behave as

$$m \sim g^{\gamma_0/\beta_0}(1 + O(g^2)) \to 0,\qquad(15.3)$$

where β_0 is the first term in the beta function controlling the vanishing of the bare coupling in the continuum limit. Since all bare quark masses are formally zero, one must address these questions in terms of a renormalization scheme at a finite cutoff.

Can one work in a mass-independent regularization scheme, where mass ratios are automatically constant? Such an approach asks that the renormalization group function $\gamma(g)$ in Eq. (15.2) be chosen to be independent of the quark species. This immediately implies the constancy of all quark mass ratios. As only the first term in the perturbative expansion of $\gamma(g)$ is universal, a mass-independent scheme is indeed an allowed procedure. However, such a procedure obscures the off-diagonal m_d effect on m_u discussed above. In particular, by forcing constancy of bare mass ratios, the ratios of physical particle masses must vary as a function of cutoff, in a manner that cancels the flow from the process discussed above. The fact that physical particle mass ratios are not just a function of quark mass ratios is shown explicitly in Section 15.3, where we observe that the combination $1 - m_{\pi_0}^2/m_{\pi_\pm}^2$ is proportional to $\frac{(m_d-m_u)^2}{(m_d+m_u)\Lambda_{qcd}}$, in the chiral limit.

From a non-perturbative point of view, having physical mass ratios vary with the cutoff seems rather peculiar; indeed, the particle masses are physical quantities that would be natural to hold fixed. And, even though a mass-independent approach is theoretically possible, there is no guarantee that any given quark mass ratio will be universal between schemes. The lattice approach itself is usually implemented with physical particle masses as input. As such it is not a mass-independent regulator, making a perturbative matching to lattice results rather subtle.

Can one simply do the perturbative matching at some high energy, say 100 GeV, where "instanton" effects are exponentially suppressed and irrelevant? This point of view has several problems. First, current lattice simulations are not done at minuscule scales and non-perturbative effects are present and substantial. Furthermore, the exponential suppression of topological effects is in the inverse coupling, which runs logarithmically with the scale. As such, the non-perturbative suppression is a power law in the scale and straightforward to estimate.

Since the eta-prime mass is expected to be of order Λ_{qcd}, we know from the previous renormalization group discussion how it depends on the bare coupling in the continuum limit

$$m_{\eta'} \propto \frac{1}{a} e^{-1/(2\beta_0 g_0^2)} g_0^{-\beta_1/\beta_0^2}. \tag{15.4}$$

While this formula indeed shows an exponential suppression in $1/g_0^2$, this is cancelled by the inverse cutoff factor in just such a way that the mass of this physical particle remains finite. The ambiguity in the quark mass splitting is controlled by the mass splitting $m_{\eta'} - m_{\pi_0}$ as well as being proportional to $m_d - m_u$. Considering $m_d = 5$ MeV at a scale of $\mu = 2$ GeV, a rough estimate of the order of the u quark mass shift is

$$\Delta m_u(\mu) \sim \left(\frac{m_{\eta'} - m_{\pi_0}}{\Lambda_{qcd}} \right) (m_d - m_u) = O(1 \text{ MeV}), \tag{15.5}$$

a number comparable to typical phenomenological estimates. This result depends on the scale μ, but that dependence is only logarithmic and given by Eq. (15.3).

A particularly important observation is that the exponent controlling the coupling constant suppression in Eq. (15.4) is substantially smaller than the classical instanton action

$$\frac{1}{2\beta_0 g_0^2} = \frac{8\pi^2}{(11 - 2n_f/3)g_0^2} \ll \frac{8\pi^2}{g_0^2}. \tag{15.6}$$

This difference arises because one needs to consider topological excitations above the quantum vacuum, not the classical. Zero modes of the Dirac operator are still responsible for the bulk of the eta prime mass, but naive semi-classical arguments strongly underestimate their effect.

15.1. The general mass term

Given the confusion over the meaning of quark masses, it is interesting to explore how two-flavor QCD behaves as these quantities are varied, including the possibility of explicit CP violation through the Θ parameter. The full theory has a rather rich phase diagram, including first and second order phase transitions, some occuring when none of the quark masses vanish.

We consider the quark fields ψ as carrying implicit isospin, color, and flavor indices. Assume as usual that the theory in the massless limit maintains the $SU(2)$ flavored chiral symmetry under

$$\psi \longrightarrow e^{i\gamma_5 \tau_\alpha \phi_\alpha/2}\psi,$$
$$\overline{\psi} \longrightarrow \overline{\psi}e^{i\gamma_5 \tau_\alpha \phi_\alpha/2}. \tag{15.7}$$

Here, τ_α represents the Pauli matrices generating isospin rotations. The angles ϕ_α are arbitrary rotation parameters.

We wish to construct the most general two-flavor mass term to add to the massless Lagrangian. This should be a dimension 3 quadratic form in the fermion fields and should transform as a singlet under Lorentz transformations. For simplicity, only consider quantities that are charge neutral as well. This leaves four candidate fields, giving, for consideration, the general form

$$m_1\overline{\psi}\psi + m_2\overline{\psi}\tau_3\psi + im_3\overline{\psi}\gamma_5\psi + im_4\overline{\psi}\gamma_5\tau_3\psi. \tag{15.8}$$

The first two terms are naturally interpreted as the average quark mass and the quark mass difference, respectively. The remaining two are less conventional. The m_3 term is connected with the CP-violating parameter of the theory. The final m_4 term has been used in conjunction with the Wilson discretization of lattice fermions, where it is referred to as a "twisted mass" [123, 124]. Its utility in that context is the ability to reduce lattice discretization errors. We will return to this term later when we discuss the effect of lattice artifacts on chiral symmetry.

These four terms are not independent. Indeed, consider the above flavored chiral rotation in the τ_3 direction, $\psi \to e^{i\phi\tau_3\gamma_5/2}\psi$. Under this, the composite fields transform as

$$\overline{\psi}\psi \longrightarrow \cos(\phi)\overline{\psi}\psi + \sin(\phi)i\overline{\psi}\gamma_5\tau_3\psi,$$
$$\overline{\psi}\tau_3\psi \longrightarrow \cos(\phi)\overline{\psi}\tau_3\psi + \sin(\phi)i\overline{\psi}\gamma_5\psi,$$
$$i\overline{\psi}\tau_3\gamma_5\psi \longrightarrow \cos(\phi)i\overline{\psi}\tau_3\gamma_5\psi - \sin(\phi)\overline{\psi}\psi,$$
$$i\overline{\psi}\gamma_5\psi \longrightarrow \cos(\phi)i\overline{\psi}\gamma_5\psi - \sin(\phi)\overline{\psi}\tau_3\psi. \tag{15.9}$$

This rotation mixes m_1 with m_4 and m_2 with m_3. Using this freedom, we can select any one of the m_i to vanish and a second to be positive.

The most common choice is to set $m_4 = 0$ and use m_1 for controlling the average quark mass. Then m_2 gives the quark mass difference, while CP violation appears in m_3. This, however, is only a convention.

An interesting alternative is the "twisted mass" scheme [123, 124]. This makes the choice $m_1 = 0$ and uses $m_4 > 0$ for the average quark mass. In this case, m_3 becomes the up-down mass difference, and m_2 becomes the CP-violating term. It is amusing to note that a mass difference in up and down quark in such a formulation involves the naively CP odd $i\overline{\psi}\gamma_5\psi$. The strong CP problem has been rotated into the smallness of the $\overline{\psi}\tau_3\psi$ term, which, under the usual conventions, is the mass difference.

Because of the flavored chiral symmetry, both sets of conventions are physically equivalent. For the following, we take the more conventional choice $m_4 = 0$, although one should keep in mind that this is only a convention and we could have chosen any of the four parameters in Eq. (15.8) to vanish.

With this choice, two-flavor QCD, after scale setting, depends on three mass parameters:

$$m_1\overline{\psi}\psi + m_2\overline{\psi}\tau_3\psi - im_3\overline{\psi}\gamma_5\psi. \tag{15.10}$$

It is the possible presence of m_3 that represents the strong CP problem. As all the parameters are independent and transform differently under the symmetries of the problem, there is no connection between the strong CP problem and m_1 or m_2.

As discussed extensively above, the chiral anomaly is responsible for the iso-singlet rotation

$$\psi \longrightarrow e^{i\gamma_5\phi/2}\psi,$$
$$\overline{\psi} \longrightarrow \overline{\psi}e^{i\gamma_5\phi/2} \tag{15.11}$$

not being a valid symmetry, despite the fact that γ_5 naively anti-commutes with the massless Dirac operator. Chapter 13 showed this anomaly is nicely summarized via Fujikawa's [97] approach where the fermion measure in the path integral picks up a non-trivial factor. In any given gauge configuration, only the zero eigenmodes of $\not{D}$ contribute, and they are connected to the winding number of the gauge configuration by the index theorem. The conclusion is that the above rotation changes the fermion measure by an amount depending non-trivially on the gauge field configuration.

Note that this anomalous rotation allows one to remove any topological term from the gauge part of the action. Naively, Θ would represent yet another parameter for the theory, but by including the general form for the fermions, this can be absorbed. Once this has been done, all that remains are the three mass parameters above, all of which are independent and relevant to physics.

These parameters are a complete set for two-flavor QCD; however, this choice differs somewhat from what is often discussed. Formally, we can define the more conventional variables as

$$m_u = m_1 + m_2 + im_3,$$

$$m_d = m_1 - m_2 + im_3,$$

$$e^{i\Theta} = \frac{m_1^2 - m_2^2 - m_3^2 + 2im_1m_3}{\sqrt{m_1^4 + m_2^4 + m_3^4 + 2m_1^2m_3^2 + 2m_2^2m_3^2 - 2m_1^2m_2^2}}. \qquad (15.12)$$

Particularly for Θ, this is a rather complicated change of variables. For non-degenerate quarks in the context of the phase diagram discussed below, the variables $\{m_1, m_2, m_3\}$ are more natural.

15.2. The strong CP problem and the up quark mass

Strong interactions preserve CP to high accuracy [125]. Thus, only two of the three possible mass parameters seem to be needed. With the above conventions, it is natural to ask: why is m_3 so small? It is the concept of unification that brings this question to the fore. We know that the weak interactions violate CP. Thus, if the electroweak and the strong interactions separate at some high scale, shouldn't some remnant of this breaking survive? How is CP recovered for the strong force?

One possible solution is that there is no unification and one should just consider the weak interactions as a small perturbation. Another approach involves adding a new dynamical "axion" field that couples to the quarks through a coupling to $i\bar{\psi}\gamma_5\psi$. Shifts in this field make m_3 essentially dynamical, and potentially, the theory could relax to $m_3 = 0$.

It is occasionally suggested that the up quark mass might vanish. This would naively allow a flavored chiral rotation to remove any phases from the quark mass matrix. However, this approach ignores the complexity of the various contributions to the quark masses. From the above discussion, one might define the up quark mass as a complex number

$$m_u \equiv m_1 + m_2 + im_3. \qquad (15.13)$$

But the quantities m_1, m_2, and m_3 are independent parameters with different symmetry properties. With our conventions, m_1 represents an isosinglet mass contribution, m_2 is isovector in nature, and m_3 is CP-violating. And, as discussed earlier, the combination $m_1 + m_2 = 0$ is scale and scheme dependent. The strong CP problem only requires small m_3. So while it may be true formally that

$$m_1 + m_2 + im_3 = 0 \implies m_3 = 0, \tag{15.14}$$

this would depend on scale and might well be regarded as "not even wrong."

15.3. The two-flavor phase diagram

As a function of the three mass parameters, QCD has a rather intricate phase diagram. Using simple chiral Lagrangian arguments, this can be qualitatively mapped out. To begin, consider composite fields similar to those used in the earlier discussion of pions as Goldstone bosons[1]

$$\sigma \sim \overline{\psi}\psi \qquad \eta \sim i\overline{\psi}\gamma_5\psi$$

$$\vec{\pi} \sim i\overline{\psi}\gamma_5\vec{\tau}\psi \qquad \vec{a}_0 \sim \overline{\psi}\vec{\tau}\psi. \tag{15.15}$$

In terms of these, a natural model for a starting effective potential is

$$V = \lambda(\sigma^2 + \vec{\pi}^2 - v^2)^2 - m_1\sigma - m_2 a_{03} - m_3\eta$$

$$+\alpha(\eta^2 + \vec{a}_0^2) - \beta(\eta\sigma + \vec{a}_0 \cdot \vec{\pi})^2. \tag{15.16}$$

Here, α and β are "low energy constants," with operators written in forms that preserve chiral symmetry. The first of these, α, gives a mass to the eta and a_0. Its presence should be regarded as a consequence of the anomaly. The β term brings in a coupling of $(\sigma, \vec{\pi})$ with $(\eta, \vec{a}_0)$. As discussed in Ref. [101], the sign of the β term is suggested so that $m_\eta < m_{a_0}$.

This potential augments the famous "Mexican hat" or "wine bottle" potential sketched earlier in Fig. 12.2. The Goldstone pions are associated with the flat directions running around at constant $\sigma^2 + \vec{\pi}^2 = v^2$. The m_2 and m_3 terms do not directly affect the σ and π fields, but induce expectation values for a_{03} and η, respectively; the size of these expectations is controlled by α. This in turn results in the β term inducing a warping of the Mexican hat into two separate minima, as sketched in Fig. 15.3. The direction of this warping is determined by the relative size of m_2 and m_3; m_2 (m_3) warps downward in π_0 (σ) direction. If we now turn on m_1, this will select one of the two minima as favored. This gives rise to a generic first order transition at $m_1 = 0$.

There is additional structure in the m_1, m_2 plane when m_3 vanishes. In this situation, the quadratic warping is downward in the π_3 direction. For large $|m_1|$, only σ will have an expectation, with sign determined by the

---

[1] What was referred to as η' at the beginning of this chapter is shortened to η here.

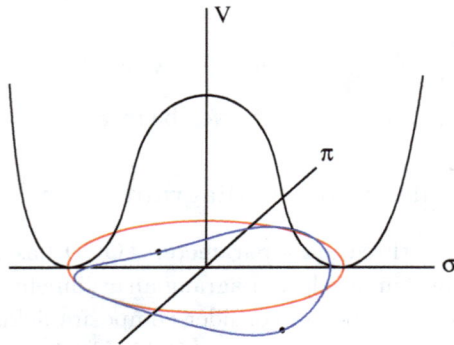

Figure 15.3:   The m_2 and m_3 terms warp the Mexican hat potential into two separate minima. The direction of the warping is determined by the relative size of these parameters. The conventional parameters $\{m_u, m_d, \Theta\}$ map non-linearly onto the tilt, the warp, and the relative angle between them. (Figure taken from Ref. [119]).

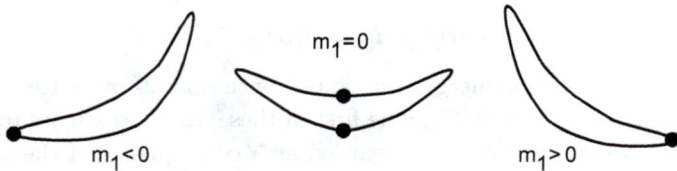

Figure 15.4:   In the m_1, m_2 plane, $m_{\pi_0}^2$ can pass through zero. This occurs when the warping overcomes the tilting to determine the minimum of the potential. This gives rise to pion condensation occuring at an Ising-like transition. (Figure taken from [101]).

sign of m_1. The pion will be massive, but with m_2 reducing the neutral pion mass below that of the charged pions. If m_1 is now decreased in magnitude at fixed m_2, the π_3 field eventually becomes massless and condenses. How this occurs is sketched in Fig. 15.4. An order parameter for the transition is the expectation value of the neutral pion field, with the transition being Ising-like.

In this simple model, the ratio of the neutral to charged pion masses can be estimated from a quadratic expansion around the minimum of the potential. For $m_3 = 0$ and m_1 above the transition line, this gives

$$\frac{m_{\pi_0}^2}{m_{\pi_\pm}^2} = 1 - \frac{\beta v m_2^2}{2\alpha^2 m_1} + O(m^2). \tag{15.17}$$

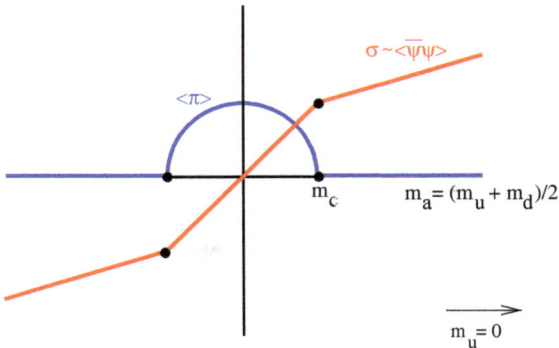

Figure 15.5:   With a constant non-vanishing up-down quark mass difference, the jump in the chiral condensate splits into two second order transitions. The order parameter distinguishing the intermediate phase is the expectation value of the neutral pion field.

The second order transition is located where this vanishes, and thus occurs for m_1 proportional to m_2^2. Note that this equation verifies the important result that a constant quark mass ratio does not correspond to a constant meson mass ratio and vice-versa. This is the ambiguity discussed at the beginning of this chapter.

The structure of this transition manifests itself in the expectation values for the pion and sigma fields as functions of the average quark mass at fixed quark mass difference, as sketched in Fig. 15.5. The jump in σ as we go from large positive to large negative masses splits into two transitions with the pion field acquiring an expectation value in the intermediate region.

The second order transition occurs in the region where both m_u and m_d are non-vanishing but of opposite relative sign. By having one quark of negative mass, one avoids the Vafa-Witten theorem [111], which says that no parity breaking phase transition can occur if the fermion determinant is positive definite.

At the transition, the correlation length diverges. This demonstrates the possibility of significant long distance physics without the presence of small Dirac eigenvalues for the Dirac operator. In contrast, we see that there is no transition at the point where only one of the quark masses vanishes. In this contrasting situation, there is no long distance physics despite the possible existence of small Dirac eigenvalues. The CP-violating phase is sketched as a function of the original up and down quark masses in Fig. 15.6.

Putting it all together, we obtain the full three-parameter phase diagram sketched in Fig. 15.7. There are two intersecting first order transition

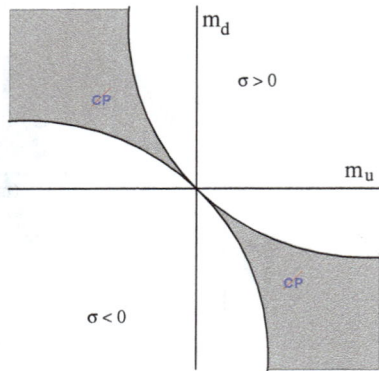

Figure 15.6:   As functions of the up and down quark masses, the CP-violating phase only occurs in the region where the quark masses differ in relative sign, but does not completely fill that region.

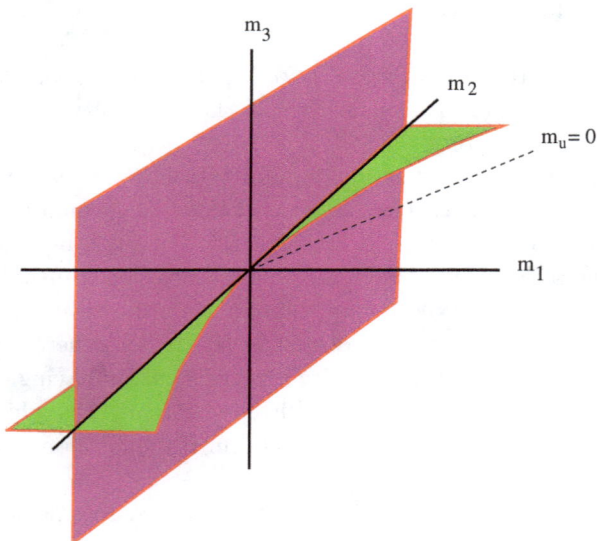

Figure 15.7:   The full phase diagram for two-flavor QCD as a function of the three mass parameters. It consists of two intersecting first order surfaces with a second order edge along curves satisfying $m_3 = 0$, $|m_1| < |m_2|$. There is no structure along the $m_u = 0$ line except when both quark masses vanish. (Figure from Ref. [119]).

surfaces, one at $(m_1 = 0, m_3 \neq 0)$ and the second at $(m_1 < m_2, m_3 = 0)$. These each occur where $\Theta = \pi$. However, note that with non-degenerate quarks, there is also a $\Theta = \pi$ region at $m_2 = m_1 + \epsilon$ for small but non-vanishing ϵ where there is no transition. The absence of a physical singularity at $m_u = 0$ when $m_d \neq 0$ lies at the heart of the problem in defining a vanishing up quark mass.

In Chapter 16, we will see that the structure in the m_1, m_2 plane is closely related to an interesting lattice artifact in the degenerate quark limit. Aoki [126] discussed a possible phase with spontaneous parity violation with the Wilson fermion formulation. Indeed, lattice artifacts can modify the effective potential in a similar way to the m_2 term and allow the CP-violating phase at finite cutoff to include part of the m_1 axis as well.

Chapter 16

Lattice fermions

We now have a coherent picture of how the spectrum of pseudo-scalar mesons is connected with chiral symmetry in the continuum theory. The anomaly plays a crucial role in introducing the Θ parameter into the theory and contributing to the η' mass. Throughout, we have assumed that we have a regulator to define the various composite fields, but have not been specific as to how that regulator is formulated. Early sections indicated the lattice should provide a natural route to a non-perturbative formulation, but we have postponed the details until some of the desired continuum features were elucidated.

The lattice should be regarded as a fully non-perturbative definition of a quantum field theory. The entire structure explored in previous chapters should emerge as we approach the continuum limit. But there are a variety of interesting and subtle issues concerning how this comes about. When the lattice is in place, all infinities in the theory are automatically removed. Yet, we have argued that the anomaly is closely tied to the divergences in the theory. As such, the associated physics must appear somewhere in any valid lattice formulation. If we try to formulate a lattice version of QCD with all classical symmetries in place, there is no way for this to work. In particular, the constraints of the anomaly impose subtleties for any valid quark action. Here, we go into this problem in some detail and explore some methods to deal with it.

16.1. Hopping and doublers

The essence of what is known as the "doubling problem" already appears in the quantum mechanics of the simplest fermion Hamiltonian in one space dimension

$$H = iK \sum_j a_{j+1}^\dagger a_j - a_j^\dagger a_{j+1}. \tag{16.1}$$

Here, j is an integer labeling the sites of an infinite chain and the a_j are fermion annihilation operators satisfying standard anti-commutation relations

$$\left[a_j, a_k^\dagger\right]_+ \equiv a_j a_k^\dagger + a_k^\dagger a_j = \delta_{j,k}. \tag{16.2}$$

The fermions hop from site to neighboring site with amplitude K; thus, we refer to K as the "hopping parameter", and by convention, take it to be positive.[1]

The bare vacuum $|0\rangle$ satisfies $a_j|0\rangle = 0$. This vacuum is not the physical one, which requires the construction of a filled Dirac sea. Energy eigenstates in the single fermion sector

$$|\chi\rangle = \sum_j \chi_j a_j^\dagger |0\rangle \tag{16.3}$$

can be easily found in momentum space

$$\chi_j(q) = e^{iqj} \chi_0 \tag{16.4}$$

where we can restrict $-\pi < q \leq \pi$. The resulting energy is

$$E(q) = 2K \sin(q). \tag{16.5}$$

The physical vacuum fills all the negative energy states, i.e. those with $-\pi < q < 0$.

On this vacuum, consider constructing a fermionic wave packet by exciting a few modes of small momentum q. This packet will have a group velocity $dE/dq \sim 2K$ that is positive. Thus, it moves to the right and represents a right-moving fermion. On the other hand, a wave packet of low energy can also be produced by exciting momenta in the vicinity of $q \sim \pi$. This packet will have group velocity $\left.\frac{dE}{dq}\right|_{q=\pi} \sim -2K$ and therefore, be left-moving. The essence of the Nielsen–Ninomiya theorem [127] is that we must have

---

[1] If K were negative, redefining $a_j \to (-1)^j a_j$ will change its sign.

both types of excitation. We will go into this in more detail later, but for this one-dimensional case, the periodicity in q requires the dispersion relation to have an equal number of zeros with positive and negative slopes. If we consider a two-component spinor to describe the fermion, we will have independent states corresponding to each component. This is the so-called "doubling" issue.

16.2. Spinors and chiral symmetry

Working our way towards a Hamiltonian version of the Wilson approach, consider a two-component spinor

$$\psi = \begin{pmatrix} a \\ b \end{pmatrix}, \tag{16.6}$$

where a and b are distinct fermion annihilation operators on the lattice sites. Remaining in one space dimension, the so called "naive" lattice Hamiltonian begins with the simple hopping case above and adds in the lower components and a mass term that mixes the two, so that

$$H = iK \sum_j a^\dagger_{j+1} a_j - a^\dagger_j a_{j+1} - b^\dagger_{j+1} b_j - b^\dagger_j b_{j+1} + M \sum_j a^\dagger_j b_j + b^\dagger_j a_j. \tag{16.7}$$

Introducing two by two Dirac matrices

$$\gamma_0 = \sigma_1 = \begin{pmatrix} 0 & 1 \\ 1 & 0 \end{pmatrix}, \quad \gamma_1 = \sigma_2 = \begin{pmatrix} 0 & -i \\ i & 0 \end{pmatrix}, \quad \gamma_5 = \sigma_3 = \begin{pmatrix} 1 & 0 \\ 0 & -1 \end{pmatrix}, \tag{16.8}$$

and defining $\overline{\psi} = \psi^\dagger \gamma_0$, we write the Hamiltonian compactly as

$$H = \sum_j K(\overline{\psi}_{j+1} \gamma_1 \psi_j - \overline{\psi}_j \gamma_1 \psi_{j+1}) + M \sum_j \overline{\psi}_j \psi_j. \tag{16.9}$$

This looks very much like the continuum Dirac Hamiltonian with the derivative term represented on the lattice by a nearest neighbor difference. Chiral symmetry is manifest in the possibility of independent rotations of the a and b type particles when the mass term is absent. The latter mixes these components and opens a gap in the spectrum.

As before, the single particle states are found by Fourier transformation and satisfy

$$E^2 = 4K^2 \sin^2(q) + M^2. \tag{16.10}$$

At each momentum, there is one positive and one negative energy state. Again, we are to fill the negative energy sea to form the physical vacuum. The doubling issue is that there are low energy excitations that satisfy the Dirac equation appearing both at $q \sim 0$ and $q \sim \pi$. The physical momenta k of the latter appear in expanding about π, i.e. $k = q - \pi$. These states have a smooth spatial dependence in a redefined field $\chi_j = (-1)^j \psi_j$. The doublers at $q \sim \pi$ are still with us.

16.3.   Wilson fermions

One way to remove the degeneracy of the doublers is to make the mixing of the upper and lower components momentum-dependent. A simple way of doing this was proposed by Wilson [13]. To our Hamiltonian model, we add one more term, so that

$$
H = iK \sum_j a^\dagger_{j+1} a_j - a^\dagger_j a_{j+1} - b^\dagger_{j+1} b_j + b^\dagger_j b_{j+1}
$$

$$
+ M \sum_j a^\dagger_j b_j + b^\dagger_j a_j - rK \sum_j a^\dagger_j b_{j+1} + b^\dagger_j a_{j+1} + b^\dagger_{j+1} a_j + a^\dagger_{j+1} b_j
$$

$$
= \sum_j K(\overline{\psi}_{j+1}(\gamma_1 - r)\psi_j - \overline{\psi}_j(\gamma_1 + r)\psi_{j+1}) + \sum_j M\overline{\psi}_j \psi_j. \qquad (16.11)
$$

Now the spectrum satisfies

$$
E^2 = 4K^2 \sin^2(q) + (M - 2rK\cos(q))^2. \qquad (16.12)
$$

The doublers at $q \sim \pi$ are increased in energy relative to the states at $q \sim 0$. The physical particle mass is now $m = M - 2rK$ while that of the doubler is at $M + 2rK$.

When r becomes large, the dip in the spectrum near $q = \pi$ actually becomes a maximum. This is irrelevant for our discussion, although we note that the case $r = 1$ is somewhat special. For this value, the matrices $(\gamma_1 \pm 1)/2$, which determine how the fermions hop along the lattice, are projection operators. In a sense, the doubler is removed because only one component can hop. This choice $r = 1$ has been the most popular in practice.

The hopping parameter has a critical value at

$$
K_c = \frac{M}{2r}. \qquad (16.13)
$$

At this point the gap in the spectrum closes, and one species of fermion becomes massless. The Wilson term, proportional to r, still mixes the a and

b type particles; so, there is no exact chiral symmetry. In the continuum limit, this represents a candidate for a chirally symmetric theory. But before the limit, chiral symmetry is broken by lattice artifacts and does not provide a good order parameter.

Now we generalize this approach to the Euclidean path integral formulation in four space-time dimensions. In the continuum, one usually writes, for the free fermion action density,

$$\overline{\psi}D\psi = \overline{\psi}(\partial\!\!\!/ + m)\psi, \tag{16.14}$$

or in momentum space,

$$\overline{\psi}(i p\!\!\!/ + m)\psi. \tag{16.15}$$

By convention we use Hermitian gamma matrices. Note that D is the sum of Hermitian and anti-Hermitian parts. In the continuum the former is just a constant, the mass. A Hermitian operator appears in the combination $\gamma_5 D$, but we do not need that right now.

A matrix can be diagonalized when it commutes with its adjoint; then it is called "normal." For the naive continuum operator this is the case, and we see that all eigenvalues of D lie along a line parallel to the imaginary axis intersecting the real axis at m. This simple structure is lost in most lattice approaches.

As discussed earlier, a simple transcription of derivatives onto the lattice replaces factors of p with trigonometric functions. Thus, the naive lattice action becomes

$$\overline{\psi}\left(\frac{i}{a}\sum_{\mu}\gamma_{\mu}\sin(p_{\mu}a) + m\right)\psi \tag{16.16}$$

where we have explicitly included the lattice spacing a. For small momentum, this reduces to the continuum result $\overline{\psi}(i\gamma_{\mu}p_{\mu} + m)\psi$. Now let one component of p get large and be near π/a. Then, we again have small eigenvalues and a nearby pole in the propagator. As any of the four components of momentum can be near 0 or π, there are a total of 16 places in momentum space that give rise to a Dirac-like behavior. The naive fermion action gives rise to 16 doublers.

The essence of the Wilson solution to doubling is the addition of a momentum-dependent mass. Wishing to maintain only nearest neighbor terms, it also involves trigonometric functions. To maintain hyper-cubic symmetry, we put in the Wilson term symmetrically for all space-time directions. For simplicity, we set the Wilson parameter r from before to unity. Explicitly for free fields, we consider the momentum space form

$$\overline{\psi} D_W \psi = \overline{\psi} \left(\frac{1}{a} \sum_\mu (i\gamma_\mu \sin(p_\mu a) + 1 - \cos(p_\mu a)) + m \right) \psi. \qquad (16.17)$$

Now for any momentum component near π, the eigenvalues are of order $1/a$ in size. Note that the lattice artifacts in the propagator start at order $p^2 a$, rather than $O(a^2)$ as for naive fermions.

The eigenvalue structure of D_W is rather interesting. The eigenvalues for the free Wilson theory occur at

$$\lambda = \pm \frac{i}{a} \sqrt{\sum_\mu \sin^2(p_\mu a)} + \frac{1}{a} \sum_\mu 1 - \cos(p_\mu a) + m. \qquad (16.18)$$

These lie on a set of "nested circles" in the complex plane, as sketched in Fig. 16.1. Note that $m \leftrightarrow -m$ is not a symmetry. Naively it would be in the continuum, but as we discussed earlier, it cannot be so in the quantum theory when one has an odd number of flavors.

Note that to obtain real eigenvalues in Eq. (16.18), each component of the momentum must be an integer multiple of π. There are actually several critical values that can give rise to massless fermions. For $m = 0, -2, -4, -6$, and -8, we have $1, 4, 6, 4$, and 1 massless species, respectively. When interactions are present these values of the mass will also be renormalized.[2] The usefulness of a continuum limit at these alternative points is unclear.

Rescaling to lattice units and restoring the hopping parameter, the Wilson fermion action, with the site indices explicit, becomes

$$D_{W\,ij} = \delta_{i,j} + K \sum_\mu (1 - \gamma_\mu)\delta_{i,j+e_\mu} + (1 + \gamma_\mu)\delta_{i,j-e_\mu}. \qquad (16.19)$$

By taking the coefficient r of the Wilson term as unity, we have projection operators in the hoppings. The physical fermion mass is read off from the small momentum behavior as $m = \frac{1}{2a}(1/K - 8)$. This vanishes at $K = K_c = 1/8$.

Here we consider that the gauge fields are formulated as usual with group valued matrices on the lattice links. These are to be inserted into the

---

[2] Actually the 6-flavor case at $m = -4$ does have a discrete symmetry that will protect against additive mass renormalization.

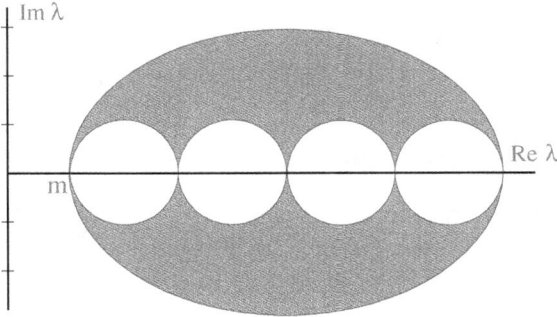

Figure 16.1: The eigenvalue spectrum of the free Wilson fermion operator is a set of nested circles. Upon turning on the gauge fields, some eigenvalues drift into the open regions. Some complex pairs can collide and become real. These are connected to gauge field topology. (Figure taken from Ref. [128]).

above hopping terms. One could use the simple Wilson gauge action as a sum over plaquettes

$$S_g = \frac{\beta}{3} \sum_p \text{Re Tr } U_p. \qquad (16.20)$$

Minor variations on this specific form should not be essential to the qualitative phase diagram. When the gauge fields are turned on, the dynamics will move the fermion eigenvalues around, partially filling the holes in eigenvalue pattern of Fig. 16.1. Complex pairs of eigenvalues can collide and become real. This is directly related to gauge field topology [112].

For the free theory, the Hermitian and anti-Hermitian parts of the action commute. This ceases to be true in the interacting case since both terms contain gauge matrices that do not themselves commute. Thus, the left eigenvalues are generally different from the right ones. Nevertheless, it is still true that the eigenvalues either appear in complex conjugate pairs or they are real. This follows from γ_5 hermiticity, $D^\dagger = \gamma_5 D \gamma_5$. Since γ_5 has unit determinant, $|D - \lambda| = 0$ implies $|D^- - \lambda| = |D - \lambda^*|^* = 0$.

A technical difficulty with this approach is that gauge interactions will renormalize the parameters. To obtain massless pions one must finely tune K to K_c, an *a priori* unknown function of the gauge coupling. Despite the awkwardness of such tuning, this is how numerical simulations with Wilson quarks generally proceed. The hopping parameter is adjusted to get the pion mass right, and one assumes that the remaining predictions of current algebra reappear in the continuum limit.

Note that the basic lattice theory has two parameters, β and K. These are related to bare coupling, $\beta \sim 6/g_0^2$, and quark mass, $(1/K - 1/K_c) \sim m_q$. We will now discuss this relation in more detail.

16.4. Lattice versus continuum parameters

As emphasized earlier, QCD is a remarkably economical theory in terms of the number of adjustable parameters. First of these is the overall strong interaction scale, Λ_{qcd}. This is scheme dependent, but once a renormalization procedure has been selected, it is well-defined. It is not independent of the coupling constant, the connection being fixed by asymptotic freedom. In addition, the theory depends on the renormalized quark masses m_i, or more precisely the dimensionless ratios m_i/Λ_{qcd}. As with the overall scale, the definition of m_i is scheme dependent. The two-flavor theory with degenerate quarks and $\Theta = 0$ has one such mass parameter. Using a lattice regulator introduces a cutoff scale given by the lattice spacing a. As with everything else, it is convenient to measure this in units of the overall scale; so, a third parameter for the cutoff theory is $a\Lambda_{qcd}$.

How the bare parameters behave as the continuum limit is taken was discussed rather abstractly in Chapter 9. The goal here is to explore some of the lattice artifacts that arise with Wilson fermions [13]. On the lattice it is generally easier to work directly with lattice parameters. One of these is the plaquette coupling β. With the usual conventions, this is related to the bare coupling $\beta = 6/g_0^2$. For the quarks, the natural lattice quantity is the "hopping parameter" K. And finally, the connection with physical scales appears via the lattice spacing a.

The set of physical parameters and the set of lattice parameters are, of course, equivalent. There is a well-understood non-linear mapping between them:

$$\left\{ a\Lambda_{qcd}, \frac{m}{\Lambda_{qcd}} \right\} \longleftrightarrow \{\beta, K\}. \qquad (16.21)$$

To extract physical predictions, we are interested in the continuum limit $a\Lambda_{qcd} \to 0$. For this, asymptotic freedom tells us we must take $\beta \to \infty$ at a rate tied to Λ_{qcd}. Simultaneously, we must take the hopping parameter to a critical value. With normal conventions, this takes $K \to K_c \to 1/8$ at a rate tied to desired quark mass m. Figure 16.2 sketches how the continuum limit is taken in the β, K plane. Next we will further explore this phase diagram with particular attention to hopping parameters larger than K_c.

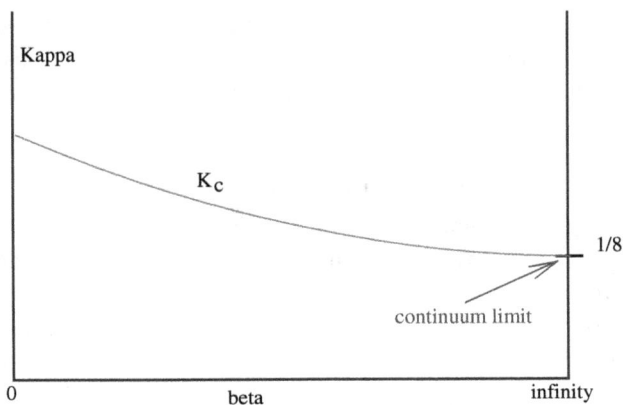

Figure 16.2: The continuum limit of lattice gauge theory with Wilson fermions occurs at $\beta \to \infty$ and $K \to 1/8$. Coming in from this point to finite beta is the curve $K_c(\beta)$, representing the lowest phase transition in K for fixed beta. The nature of this phase transition is a delicate matter, discussed in the text. (Figure taken from Ref. [128]).

This discussion is adapted from Ref. [128] and adds the possible twisted mass term to the presentation in Ref. [129].

16.5. Artifacts and the Aoki phase

We previously made extensive use of an effective field theory to describe the interactions of the pseudo-scalar mesons. Here we will begin with the simplest form for the two-flavor theory and then add terms to mimic possible lattice artifacts. The language is framed as before in terms of the isovector pion field $\vec{\pi} \sim i\bar{\psi}\gamma_5\vec{\tau}\psi$ and the scalar sigma $\sigma \sim \bar{\psi}\psi$. The starting point for this discussion is the canonical "Mexican hat" potential

$$V_0 = \lambda \left(\sigma^2 + \vec{\pi}^2 - v^2\right)^2 \tag{16.22}$$

schematically sketched earlier in Fig. 12.2. The potential has a symmetry under $O(4)$ rotations between the pion and sigma fields expressed as the four-vector $\Sigma = (\sigma, \vec{\pi})$. This represents the axial symmetry of the underlying quark theory.

As discussed before, the massless theory is expected to spontaneously break chiral symmetry. This is due to the minimum energy for the potential occurring at a non-vanishing value for the fields. As usual, we take the vacuum to lie in the sigma direction with $\langle\sigma\rangle > 0$. The pions are then

Goldstone bosons, being massless because the potential provides no barrier to oscillations of the fields in the pion directions. Also as discussed before, we include a quark mass by adding a constant times the sigma field

$$V_1 = -m \ \sigma. \tag{16.23}$$

This explicitly breaks the chiral symmetry by "tilting" the potential, as sketched in Fig. 12.3, hence selecting a unique vacuum which, for $m > 0$, gives a positive expectation for sigma. In the process the pions gain a mass, with $m_\pi^2 \sim m$.

Because of the symmetry of V_0, it does not matter physically which direction we tilt the vacuum in. In particular, a mass term of form

$$m \ \sigma \rightarrow m \ \sigma \cos(\theta) + m\pi_3 \sin(\theta) \tag{16.24}$$

should give equivalent physics for any θ. In the earlier continuum discussion, we used this freedom to rotate the second term away. However, as we will see, lattice artifacts can break the symmetry, introducing the possibility of physics at finite lattice spacing depending on this angle. As mentioned before, the second term in this equation is what is usually called a "twisted mass."

The Wilson term inherently breaks chiral symmetry. This will give rise to various modifications of the effective potential. The first correction is expected to be an additive contribution to the quark mass, i.e. an additional tilt to the potential. This means that the critical kappa, defined as the smallest kappa where a singularity is found in the β, K plane, will move away from the limiting value of $1/8$. Thus, we introduce the function $K_c(\beta)$ and imagine that the mass term is modeled with the form

$$m \rightarrow c_1 \ (1/K - 1/K_c(\beta)), \tag{16.25}$$

with c_1 a constant parameter.

In general, lattice modifications of the effective potential will introduce further corrections that go away in the continuum limit. It is natural to model such corrections as an expansion in the chiral fields. With this motivation, we include a term in the potential of form $c_2 \ \sigma^2$ and generalize the effective model to

$$V(\vec{\pi}, \sigma) = \lambda \ (\sigma^2 + \vec{\pi}^2 - v^2)^2 - c_1 \ (1/K - 1/K_c(\beta)) \ \sigma + c_2 \ \sigma^2. \tag{16.26}$$

Such a term was considered in Refs. [129, 130]. The predicted phase structure depends qualitatively on the sign of c_2, but, a priori, we have

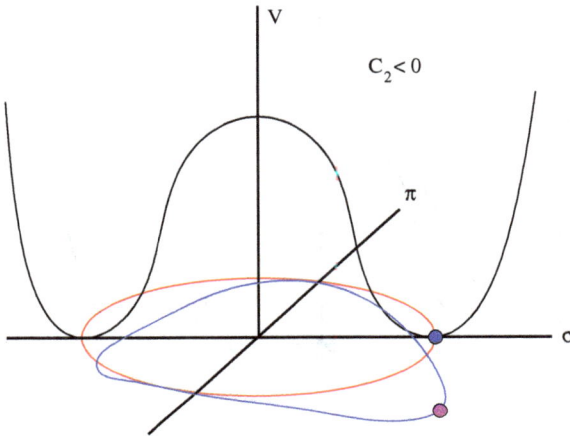

Figure 16.3: Lattice artifacts could quadratically warp the effective potential. If this warping is downward in the sigma direction, the chiral transition becomes first order without the pions becoming massless. (Figure taken from Ref. [128]).

no information on this.[3] Indeed, as it is a lattice artifact, it is expected that this sign might depend on the choice of gauge action. Note that we could have added a term proportional to $\vec{\pi}^2$, but this is essentially equivalent since $\vec{\pi}^2 = (\sigma^2 + \vec{\pi}^2) - \sigma^2$, and the first term here can be absorbed, up to an irrelevant constant, into the Mexican hat potential we started with.

First consider the case when c_2 is less than zero, thus lowering the potential energy when the field points in the positive or negative sigma direction. This quadratic warping helps to stabilize the sigma direction, as sketched in Fig. 16.3, and the pions cease to be true Goldstone bosons when the quark mass vanishes. Instead, as the mass passes through zero, we have a first order transition as the expectation of σ jumps from positive to negative. This jump occurs without any physical particles becoming massless.

Things get a bit more complicated if $c_2 > 0$, as sketched in Fig. 16.4. In that case, the chiral transition splits into two second order transitions separated by a phase with an expectation for the pion field, i.e. $\langle \vec{\pi} \rangle \neq 0$. The behavior is directly analogous to that shown in Fig. 15.5, the main difference being that the two quarks are now degenerate. Since the pion field has odd parity and charge conjugation and carries isospin as well, all of these symmetries are spontaneously broken in the intermediate phase. As isospin

---

[3]Reference [131] argues that c_2 should be positive. We will return to this argument a bit later in this chapter.

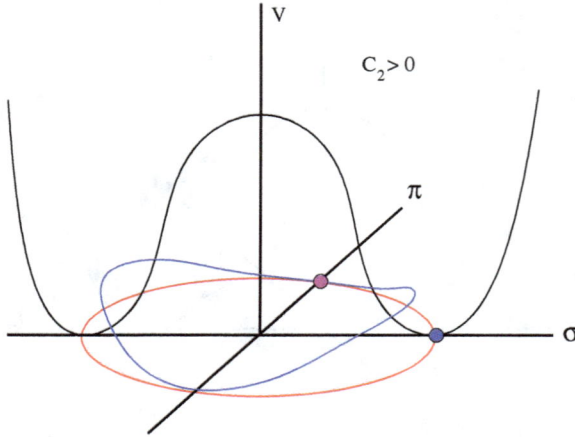

Figure 16.4: If the lattice artifacts warp the potential upward in the sigma direction, the chiral transition splits into two second order transitions separated by a phase where the pion field has an expectation value. (Figure taken from Ref. [128]).

is a continuous group, this phase will exhibit Goldstone bosons. The number of these is two, representing the two flavor generators orthogonal to the direction of the expectation value. If higher order terms do not change the order of the transitions, there will be a third massless particle exactly at the transition endpoints. In this way, the theory reveals three massless pions exactly at the transitions, as discussed by Aoki [126]. The intermediate phase is usually referred to as the "Aoki phase." Assuming this $c_2 > 0$ case, Fig. 16.5 shows the qualitative phase diagram expected.

This discussion has a close similarity to the treatment that led to the phase diagram in Fig. 15.7. Indeed, lattice artifacts can cause the spontaneously broken CP region to open up and remain present even for degenerate quarks. This modifies what we saw in Fig. 15.6 to the picture sketched in Fig. 16.6. The Aoki phase is thus closely related to the possibility of CP violation at $\Theta = \pi$ for unequal mass quarks.

This picture will be drastically modified with an odd number of degenerate flavors. The earlier continuum discussion showed how a spontaneous breaking of parity is the normal expectation when the mass is negative. That corresponds to the hopping parameter exceeding its critical value. In this situation, the Aoki phase is less a lattice artifact than a direct consequence of the CP violation expected in the continuum theory at $\theta = \pi$.

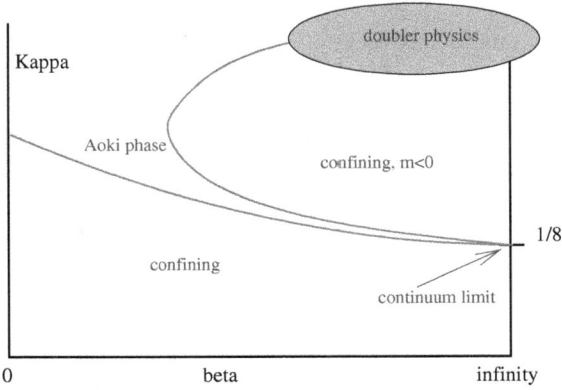

Figure 16.5: The qualitative structure of the β, K plane, including the possibility of an Aoki phase.

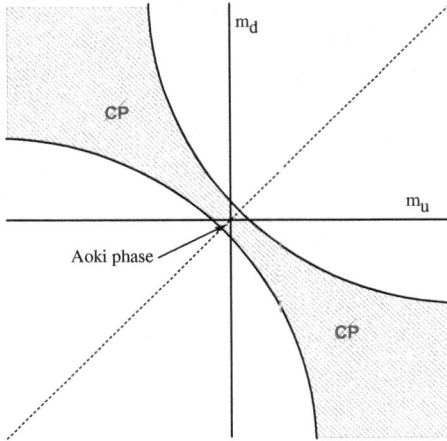

Figure 16.6: The Aoki phase opens the CP-violating phase to include a region where the quarks are degenerate.

16.6. Twisted mass

The c_2 term breaks the equivalence of different chiral directions. This means that physics will indeed depend on the angle θ if one takes a mass term of the form in Eq. (16.24). Consider complexifying the fermion mass in the usual way

$$m\overline{\psi}\psi \rightarrow m\overline{\psi}_L\psi_R + m^*\overline{\psi}_R\psi_L = \operatorname{Re} m \,\overline{\psi}\psi + i\operatorname{Im} m \,\overline{\psi}\gamma_5\psi. \qquad (16.27)$$

The rotation of Eq. (16.24) is equivalent to giving the up and down quark masses opposite phases

$$m_u \rightarrow e^{+i\theta} m_u,$$
$$m_d \rightarrow e^{-i\theta} m_d. \tag{16.28}$$

Thus motivated, we can consider adding a new mass term to the lattice theory

$$\mu \, i\bar{\psi}\tau_3\gamma_5\psi \sim \mu\pi_3. \tag{16.29}$$

This extends our effective potential to

$$V(\vec{\pi}, \sigma) = \lambda(\sigma^2 + \vec{\pi}^2 - v^2)^2 - c_1(1/K - 1/K_c(\beta))\sigma + c_2\sigma^2 - \mu\pi_3. \tag{16.30}$$

The twisted mass represents the addition of a "magnetic field" conjugate to the order parameter for the Aoki phase.

There are a variety of motivations for adding such a term to our lattice action [132, 133]. Primary among them is that $O(a)$ lattice artifacts can be arranged to cancel. With two flavors of conventional Wilson fermions, these effects change sign on going from positive to negative mass. If we put all the mass into the twisted term we are halfway between. It should be noted that this cancellation only occurs when all the mass comes from the twisted term; for other combinations with a traditional mass term, some lattice artifacts of $O(a)$ will survive. Also, although it looks like we are putting phases into the quark masses, these cancel between the two flavors. The resulting fermion determinant remains positive and we are working where the physical CP-violating parameter $\Theta = 0$. Furthermore, the algorithm is considerably simpler and faster than either overlap [100, 134] or domain wall [135, 136] fermions while avoiding the diseases of staggered quarks [137].[4] Another nice feature of adding a twisted mass term is that it allows a better understanding of the Aoki phase and shows how to continue around it. Figures 16.7 and 16.8 show how this works for the case $c_2 > 0$ and $c_2 < 0$, respectively.

Of course some difficulties come along with these advantages. First, one needs to know K_c. Indeed, with the Aoki phase present, the definition of this quantity is a bit fuzzy. And second, the mass needs to be larger than the c_2 artifacts. Indeed, as Figs. 16.7 and 16.8 suggest, if it is not, then one

---

[4]These approaches will be discussed in the next chapter.

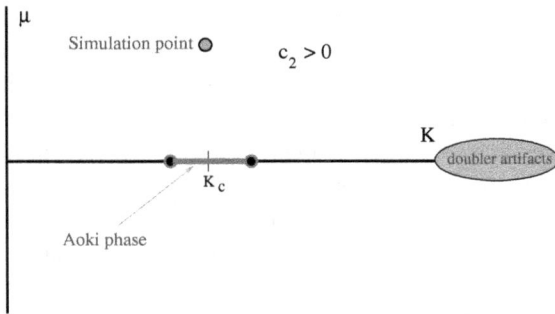

Figure 16.7: Continuing around the Aoki phase with twisted mass. This sketch considers the case $c_2 > 0$ where the parity broken phase extends over a region along the kappa axis. (Figure taken from Ref. [128]).

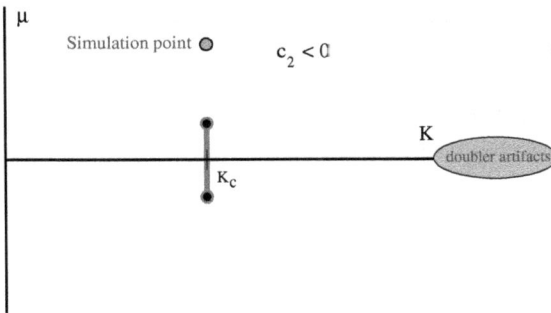

Figure 16.8: As in Fig. 16.7, but for the case $c_2 < 0$ so the chiral transition on the kappa axis becomes first order. (Figure taken from Ref. [128]).

is really studying the physics of the Aoki phase, not the continuum limit. This also has implications for how close to the continuum one must be to study this structure; in particular, one must have β large enough so the Aoki phase does not extend into the doubler region.

The question of the sign of c_2 remains open. Simulations suggest that the usual Aoki phase with $c_2 > 0$ is the situation with the Wilson gauge action. Ref. [131] showed that with the twisted mass term present, all eigenvalues of the product of gamma five with the Dirac operator will have non-zero imaginary parts. Thus to have $c_2 < 0$, the phase transition at non-vanishing twisted mass must occur where the fermion determinant does not vanish on any configuration. This contrasts with the $c_2 > 0$ case where small eigenvalues of D are expected in the vicinity of the critical hopping. This, at first sight, makes $c_2 < 0$ seem somewhat unnatural; however, this is not a proof since we saw in Section 15.3 that phase transitions without small eigenvalues of the Dirac operator can occur in the continuum theory with non-degenerate quarks.

This picture of the artifacts associated with Wilson fermions raises some interesting questions. One concerns possible generalizations of twisted mass to the three-flavor theory. As mentioned before, a negative mass represents the parity broken phase at $\Theta = \pi$. Twisting needs to be done in a way that avoids this region. Also, the twisting process is not unique, with possible twists in the λ_3 or λ_8 directions. For one example, using only λ_3 would suggest a possible twisted mass of form $m_u \sim e^{2\pi i/3}$, $m_d \sim e^{-2\pi i/3}$, $m_s \sim 1$. Whether there is an optimum twisting procedure for three flavors is unclear.

Another special case is one-flavor QCD [112]. In this situation the anomaly removes all chiral symmetry, and the quark condensate loses meaning as an order parameter. The critical value of kappa where the mass gap disappears is decoupled from the point of zero physical quark mass. A parity broken phase does exist, but it occurs only at sufficiently negative quark mass. And from the point of view of twisting the mass, without chiral symmetry there is nothing to twist other than turning on the physical parameter Θ.

Chapter 17

Actions preserving chiral symmetry

17.1. The Nielsen–Ninomiya theorem

As discussed some time ago [127], the doubling issue is closely tied to topology in momentum space. To see how this works, let us first establish a gamma matrix convention

$$\vec{\gamma} = \sigma_1 \otimes \vec{\sigma} = \begin{pmatrix} 0 & \vec{\sigma} \\ \vec{\sigma} & 0 \end{pmatrix}, \tag{17.1}$$

$$\gamma_0 = \sigma_2 \otimes I = \begin{pmatrix} 0 & -i \\ i & 0 \end{pmatrix}, \tag{17.2}$$

$$\gamma_5 = \sigma_3 \otimes I = \begin{pmatrix} 1 & 0 \\ 0 & -1 \end{pmatrix}. \tag{17.3}$$

Now, suppose we have an anti-Hermitian Dirac operator D that anti-commutes with γ_5

$$D = -D^\dagger = -\gamma_5 D \gamma_5. \tag{17.4}$$

Considering this quantity in momentum space, its most general form is

$$D(p) = \begin{pmatrix} 0 & z(p) \\ -z^*(p) & 0 \end{pmatrix} \tag{17.5}$$

where $z(p)$ is of the form

$$z(p) = z_0(p) + i\vec{\sigma} \cdot \vec{z}(p). \tag{17.6}$$

W. R. Hamilton in 1843 referred to this structure as a quaternion. Thus, we see that any chirally symmetric Dirac operator maps momentum space onto the space of quaternions.

The Dirac equation is obtained by expanding the momentum space operator around a zero, i.e. $D(p) \simeq i\not{p} = i\gamma_\mu p_\mu$. Now consider a three-dimensional sphere embedded in four-dimensional momentum-space, surrounding the zero with a constant $D^2 \sim p^2$. The above quaternion wraps non-trivially about the origin as we cover this sphere. Here is where the topology comes in [127, 138]. Momentum space is periodic over Brillouin zones. We must identify p with $p + 2\pi n$ where n is an arbitrary four vector with integer components. This allows us to restrict the momentum components to lie in the range $-\pi < p_\mu \leq \pi$. There is no real boundary on the resulting torus; we cannot have any non-trivial topology surviving on the surface of this zone. Any mapping associated with a zero in $z(p)$ must unwrap somewhere else before reaching the surface. Assuming $D(p)$ remains finite, any zero must be accompanied by another wrapping in the opposite sense. In the case of naive fermions, the 16 species pair up into 8 zeros of each sense.

The above argument only tells us that a chiral lattice theory must have an even number of species. The case of a minimal doubling with only two species is in fact possible, although all methods presented so far [139] appear to involve a breaking of hyper-cubic symmetry. The breaking is associated with the direction between the zeros; this makes one direction special, although it might be possible to avoid it by having the zeros form a symmetric lattice using the periodicity of momentum space. This has not yet been demonstrated.

In earlier chapters we discussed how an odd number of flavors raised some interesting issues; in particular the sign of the mass becomes relevant. In spite of this, there seems to be no obvious physical contradiction with having, say, three light flavors in the continuum with a well-defined chiral limit. However, the above topological argument seems to suggest troubles with maintaining an exact chiral symmetry on a lattice with an odd number of flavors. Whether this apparent conflict is serious is unclear. One could always start with a multiple fermion theory and then, with something like a Wilson term, give mass to a few species while leaving behind an odd number of massless fermions. This will involve some parameter tuning, but can presumably give a reasonable chiral limit for odd $N_f > 2$. This does not obviate the fact emphasized earlier, that with only one flavor, there must not be any remaining chiral symmetry, even in the continuum.

17.2.   Minimal doubling

Several chiral lattice actions satisfying the minimal condition of $N_f = 2$ flavors are known. Some time ago, Karsten [140] presented a simple form by inserting a factor of $i\gamma_4$ into a Wilson-like term for space-like hoppings. A slight variation appeared in a discussion by Wilczek [141] a few years later. More recently, a new four-dimensional action was motivated by the analogy with two-dimensional graphene [138]. Since then, numerous variations have been presented [139, 142–145]. An extensive review of the subject appears in J. Weber's thesis [146].

The main potential advantage with such approaches lies in ultra-locality. They involve only nearby neighbor hoppings for the fermions. Therefore, they should be extremely fast in simulations while still protecting masses from additive renormalization as well as controlling mixing of operators with different chirality. The approach also avoids the uncontrolled errors associated with the rooting approximation, as will be discussed later [83, 137, 147, 148]. On the other hand, all minimally-doubled actions presented so far have the disadvantage of breaking lattice hyper-cubic symmetry. With interactions, this leads to the necessity of renormalization counter-terms that also violate this symmetry [149]. The extent to which this will complicate practical simulations remains to be investigated.

Minimally-doubled chiral fermions have the unusual property of a single local field creating two distinct fermionic species. Here, we discuss a point-splitting method for separating the effects of the two flavors which can be created by a single fermion field. We will work with one specific form for the fermion action here, but the method is easily extended to other minimally-doubled formulations.

We concentrate on a minimally-doubled fermion action, which is a slight generalization of those presented by Karsten [140] and Wilczek [141]. The fermion term in the lattice action takes the form $\overline{\psi}D\psi$. For free fermions, we start in momentum space with

$$
D(p) = i \sum_{i=1}^{3} \gamma_i \sin(p_i) + \frac{i\gamma_4}{\sin(\alpha)} \left(\cos(\alpha) + 3 - \sum_{\mu=1}^{4} \cos(p_\mu) \right).
$$

(17.7)

The Wilson-like term for the space-like hoppings contains an extra factor of $i\gamma_4$. As a function of the momentum p_μ, the propagator $D^{-1}(p)$ has two poles, located at $\vec{p} = 0$, $p_4 = \pm\alpha$. Relative to the naive fermion action, the other doublers have been given a large "imaginary chemical potential" by the Wilson-like term. The parameter α allows the adjustment of the relative positions of the poles. The original Karsten/Wilczek actions correspond to $\alpha = \pi/2$.

This action maintains one exact chiral symmetry, manifested in the anti-commutation relation $[D, \gamma_5]_+ = 0$. The two species, however, are not equivalent, but have opposite chirality. To see this, expand the propagator around the two poles and observe that one species, that corresponding to $p_4 = +\alpha$, uses the usual gamma matrices, but the second pole gives a proper Dirac behavior using another set of matrices $\gamma'_\mu = \Gamma^{-1}\gamma_\mu\Gamma$ where $\Gamma = i\gamma_4\gamma_5$. Other minimally-doubled actions generally involve a different transformation. This second set of matrices have $\gamma'_5 = -\gamma_5$, showing that the two species rotate oppositely under the exact chiral symmetry. Because of this sign change, the full chiral symmetry should be regarded as "flavored." One might think of the physical chiral symmetry as that generated in the continuum theory by $\tau_3\gamma_5$.

It is straightforward to transform the momentum space action in Eq. (17.7) to position space and insert gauge fields $U_{ij} = U_{ji}^\dagger$ on the links connecting lattice sites. Explicitly indicating the site indices, the Dirac operator becomes

$$D_{ij} = U_{ij} \sum_{\mu=1}^{3} \gamma_i \frac{\delta_{i,j+e_\mu} - \delta_{i,j-e_\mu}}{2}$$

$$+ \frac{i\gamma_4}{\sin(\alpha)} \left((\cos(\alpha) + 3)\delta_{ij} - U_{ij} \sum_{\mu=1}^{4} \frac{\delta_{i,j+e_\mu} + \delta_{i,j-e_\mu}}{2} \right). \tag{17.8}$$

Again we see analogy with Wilson fermions [13] for the space directions, but augmented with an $i\gamma_4$ inserted in the Wilson term.

Perturbative calculations [149] have shown that interactions with the gauge fields can shift the relative positions of the poles along the direction between them. In other words, the parameter α receives an additive renormalization. Furthermore, the action treats the fourth direction differently from the spatial coordinates; this is the breaking of hyper-cubic symmetry mentioned above. There arise three possible new counter-terms for the renormalization of the theory. Firstly, there is a possible renormalization of the on-site contribution to the action proportional to $i\bar{\psi}\gamma_4\psi$. Adjusting the size of this term provides a handle on the shift of the parameter α. Secondly, the breaking of the hyper-cubic symmetry indicates that one may need to adjust the fermion "speed of light." That can be adjusted with a combination of the above on-site term and the strength of temporal hopping proportional to $\delta_{i,j+e_4} + \delta_{i,j-e_4}$. Finally, the breaking of hyper-cubic symmetry can feed back into the gluonic sector, suggesting a possible counter-term of form $F_{4\mu}F_{4\mu}$ to maintain the gluon "speed of light." In lattice language, this corresponds to adjusting the strength of time-like plaquettes relative to space-like ones.

Of these counter-terms, $i\bar{\psi}\gamma_4\psi$ is of dimension 3 and is probably the most essential. Quantum corrections induce the dimension 4 terms, suggesting they may be both small and absorbed partially by accepting a lattice asymmetry. Preliminary studies [146] suggest that these renormalizations can indeed be controlled. Note that all other dimension 3 counter-terms are forbidden by basic symmetries. For example, chiral symmetry forbids $\bar{\psi}\psi$ and $i\bar{\psi}\gamma_5\psi$ terms, and spatial cubic symmetry removes $\bar{\psi}\gamma_i\psi$, $\bar{\psi}\gamma_i\gamma_5\psi$, and $\bar{\psi}\sigma_{ij}\psi$ terms. Finally, commutation with γ_4 plus space inversion eliminates $\bar{\psi}\gamma_4\gamma_5\psi$.

The fundamental field ψ can create either of the two species. For a quantity that creates only one of them, it is natural to combine fields on nearby sites in such a way as to cancel the other. In other words, one can point split the fields to separate the poles which occur at distinct "bare momenta." For the free theory, one construction that accomplishes this is to consider

$$u(q) = \frac{1}{2}\left(1 + \frac{\sin(q_4 + \alpha)}{\sin(\alpha)}\right)\psi(q + \alpha e_4),$$

$$d(q) = \frac{1}{2}\Gamma\left(1 - \frac{\sin(q_4 - \alpha)}{\sin(\alpha)}\right)\psi(q - \alpha e_4), \tag{17.9}$$

where $\Gamma = i\gamma_4\gamma_5$ from above. Here we have inserted factors containing zeros canceling the undesired pole. This construction is not unique, and specific details will depend on the particular minimally-doubled action in use. The factor of Γ inserted in the d quark field accounts for the fact that the two species use different gamma matrices. This is required since the chiral symmetry is flavored, corresponding to an effective minus sign in γ_5 for one of the species.

Proceeding to position space and inserting link factors to keep gauge transformations simple,

$$u_x = \frac{1}{2}e^{i\alpha x_4}\left(\psi_x + i\frac{U_{x,x-e_4}\psi_{x-e_4} - U_{x,x+e_4}\psi_{x+e_4}}{2\sin(\alpha)}\right),$$

$$d_x = \frac{1}{2}\Gamma e^{-i\alpha x_4}\left(\psi_x - i\frac{U_{x,x-e_4}\psi_{x-e_4} - U_{x,x+e_4}\psi_{x+e_4}}{2\sin(\alpha)}\right). \tag{17.10}$$

The phase factors serve to remove the oscillations associated with the bare fields having their poles at non-zero momentum.

Given the basic fields for the individual quarks, one can go on to construct mesonic fields, which then also involve point splitting. To keep the equations simpler, we now restrict ourselves to the case $\alpha = \pi/2$. For example, the neutral pion field becomes

$$
\pi_0(x) = \frac{i}{2}(\overline{u}_x\gamma_5 u_x - \overline{d}_x\gamma_5 d_x)
$$

$$
= \frac{i}{16}(4\overline{\psi}_x\gamma_5\psi_x + \overline{\psi}_{x-e_4}\gamma_5\psi_{x-e_4} + \overline{\psi}_{x+e_4}\gamma_5\psi_{x+e_4}
$$

$$
-\overline{\psi}_{x+e_4}UU\gamma_5\psi_{x-e_4} - \overline{\psi}_{x-e_4}UU\gamma_5\psi_{x+e_4}). \qquad (17.11)
$$

Note that this involves combinations of fields at sites separated by either 0 or 2 lattice spacings. In contrast, the η' takes the form

$$
\eta'(x) = \frac{i}{2}(\overline{u}_x\gamma_5 u_x + \overline{d}_x\gamma_5 d_x)
$$

$$
= \frac{1}{8}(\overline{\psi}_{x-e_4}U\gamma_5\psi_x - \overline{\psi}_x U\gamma_5\psi_{x-e_4} - \overline{\psi}_{x+e_4}U\gamma_5\psi_x + \overline{\psi}_x U\gamma_5\psi_{x+e_4})
$$

$$
(17.12)
$$

where all terms connect even with odd parity sites. Tiburzi [150] has discussed how the anomaly, which gives the η' a mass of order Λ_{qcd}, can be understood in terms of the necessary point splitting.

17.3. Domain wall and overlap fermions

The overlap fermion was originally developed [151] as a limit of a fermion formulation using four-dimensional surface modes on a five-dimensional lattice. This effectively amounts to Shockley surface states [152] as the basis for a theory maintaining chiral symmetry [135, 153]. For a review, see Ref. [154]. The idea is to set up a theory in one extra dimension so that surface modes exist, and our observed world is an interface with our quarks and leptons being these surface modes. Particle hole symmetry naturally gives the basic fermions zero mass. In the continuum limit, the extra dimension becomes unobservable due to states in the interior requiring a large energy to create. In this picture, opposing surfaces carry states of opposite helicity, and the anomalies are due to a tunneling through the extra dimension [155].

Ref. [155] discusses the general conditions for surface modes to exist. Normalized solutions are bound to any interface separating a region with super-critical hopping from that with sub-critical hopping. Kaplan's original paper [135] considered not a surface, but an interface with $M = M_{cr} + m\epsilon(x)$, where M_{cr} is the critical value for the mass parameter where the five-dimensional fermions would be massless. Shamir [156] presented a somewhat simpler picture with the hopping vanishing on one side; this reduces the interface to a surface.

To couple gluon fields to this theory without adding unneeded degrees of freedom, the gauge fields are taken to lie in the four physical space-time directions and are independent of the fifth coordinate. In this approach, the extra dimension is perhaps best thought of as a flavor space [157]. With a finite lattice, this procedure gives equal couplings of the gauge field to the fermion modes on opposing walls in the extra dimension. Since the left and right handed modes are separated by the extra dimension, they only couple through the gauge field. The result is an effective light Dirac fermion. In the case of the strong interactions, this provides an elegant scheme for a natural chiral symmetry without the tuning inherent in the Wilson approach. A small breaking of chiral symmetry arises only through finiteness of the extra dimension.[1]

The name "overlap operator" comes from the overlap of eigenstates of the different five-dimensional transfer matrices on each side of the interface. Although originally derived from an infinite limit of the five-dimensional formalism, one can formulate the overlap operator directly in four dimensions. We begin with the fermionic part of some generic action as a quadratic form $S_f = \overline{\psi} D \psi$. The usual "continuum" Dirac operator $D = \sum \gamma_\mu D_\mu$ naively anti-commutes with γ_5, i.e. $[\gamma_5, D]_+ = 0$. Then the change of variables $\psi \rightarrow e^{i\theta\gamma_5}\psi$ and $\overline{\psi} \rightarrow \overline{\psi}e^{i\theta\gamma_5}$ would naively be a symmetry of the action. This, however, is inconsistent with the chiral anomalies. The conventional continuum discussion, as discussed in Chapter 13, maps this phenomenon into the fermionic measure [97].

On the lattice we work with a finite number of degrees of freedom; thus, the above variable change is automatically a symmetry of the measure. To parallel the continuum discussion, it is necessary to modify the symmetry transformation on the action so that the measure is no longer invariant. Remarkably, it is possible to construct lattice Dirac operators satisfying a modified symmetry under which corresponding actions are exactly invariant.

---

[1]The anomaly, however, shows that some communication between the surfaces survives even as the extra dimension becomes infinite. This is possible since the same gauge fields are on each surface.

To be specific, one particular variation [100, 158–161] modifies the change of variables to

$$\psi \longrightarrow e^{i\theta\gamma_5}\psi$$
$$\overline{\psi} \longrightarrow \overline{\psi}e^{i\theta(1-aD)\gamma_5} \tag{17.13}$$

where a represents the lattice spacing. Note the asymmetric way in which the independent Grassmann variables ψ and $\overline{\psi}$ are treated. Requiring the action to be unchanged gives the relation [99, 162, 163]

$$D\gamma_5 = -\gamma_5 D + aD\gamma_5 D = -\hat{\gamma}_5 D \tag{17.14}$$

with $\hat{\gamma}_5 = (1 - aD)\gamma_5$. To proceed, we also assume the Hermiticity condition $\gamma_5 D\gamma_5 = D^\dagger$. The naive anti-commutation relation of D with γ_5 receives a correction formally proportional to the lattice spacing. The above "Ginsparg-Wilson relation" along with the Hermiticity condition is equivalent to the unitarity of the combination $V = 1 - aD$.

Neuberger [158, 159] and Chiu and Zenkin [160] presented an explicit operator with the above properties. They first construct V via a unitarization of an undoubled chiral violating Dirac operator, such as the Wilson operator D_w. This operator should also satisfy the above Hermiticity condition $\gamma_5 D_w \gamma_5 = D_w^\dagger$. Specifically, they consider

$$V = -D_w(D_w^\dagger D_w)^{-1/2}. \tag{17.15}$$

The combination $(D_w^\dagger D_w)^{-1/2}$ is formally defined by finding a unitary operator to diagonalize the Hermitian combination $D_w^\dagger D_w$, taking the square root of the eigenvalues, and then undoing the unitary transformation. This process is tedious, dominating the computational effort with algorithms of this type. Note also that this process is not "ultra-local." Even if the kernel D_W is highly sparse, that will not be the case for V.

Directly from V, we construct the overlap operator as

$$D = (1 - V)/a. \tag{17.16}$$

The Ginsparg-Wilson relation of Eq. (17.14) is most succinctly written as the unitarity of V coupled with its γ_5 Hermiticity

$$\gamma_5 V\gamma_5 V = 1. \tag{17.17}$$

The basic projection process is illustrated in Fig. 17.1.

The overlap operator has several nice properties. Being constructed from a unitary operator, the normality of D is guaranteed. But, most importantly, it exhibits a lattice version of an exact chiral symmetry [165].

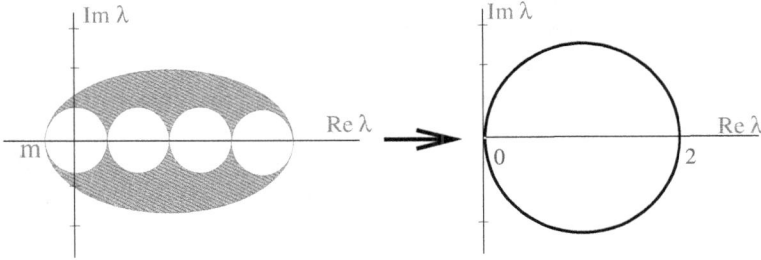

Figure 17.1: The overlap operator is constructed by projecting the eigenvalues of the Wilson operator onto a circle. (Figure taken from Ref. [164]).

The fermionic action $\overline{\psi}D\psi$ is invariant under the transformation of equation relations, i.e.

$$\psi \rightarrow e^{i\theta\gamma_5}\psi,$$
$$\overline{\psi} \rightarrow \overline{\psi}e^{i\theta\hat{\gamma}_5}, \tag{17.18}$$

where

$$\hat{\gamma}_5 = V\gamma_5. \tag{17.19}$$

As with γ_5, this quantity $\hat{\gamma}_5$ is Hermitian and its square is unity. Thus, its eigenvalues are all plus or minus unity. The trace defines an index

$$\nu = \frac{1}{2}\mathrm{Tr}\hat{\gamma}_5 \tag{17.20}$$

which plays exactly the role of the index in the continuum. If the gauge fields are smooth, this counts the topology of the gauge configuration. The factor of $1/2$ in Eq. (17.20) appears because the exact zero modes of the overlap operator have partners on the opposite side of the unitarity circle which also contribute to the trace.

At this point, the hopping parameter in D_w is an arbitrary constant. To have the desired single light fermion per flavor of the theory, the hopping parameter should be appropriately adjusted to lie above the critical value where D_w describes a massless flavor, but not so large that additional doublers come into play [166]. There are actually two parameters to play with: the hopping parameter and the lattice spacing. When the latter is finite and gauge fields are present, the location of the critical hopping parameter in D_w is expected to shift from that of the free fermion theory. As we saw when discussing the Aoki phase, there is potentially a rather complex phase structure in the plane of these two parameters, with various numbers of doublers becoming massless as the hopping is varied. The Ginsparg-Wilson relation in and of itself does not, in general, determine the number of physical massless fermions.

Although the Wilson operator entering this construction is local and quite sparse, the resulting overlap action is not. The inversion in Eq. (17.15) brings in direct couplings between arbitrarily separated sites [167–169]. How rapidly these couplings fall with distance depends on the gauge fields and is not fully understood. The five-dimensional domain-wall theory is local in the most naive sense; all terms in the action only couple nearest neighbor sites. However, were one to integrate out the heavy modes, the resulting low energy effective theory would also involve couplings with arbitrary range. Despite these non-localities, encouraging studies [159, 170–173] show that it may indeed be practical to implement the required inversion in large scale numerical simulations. The overlap operator should have memory advantages over the domain wall approach since a large number of fields corresponding to the extra dimension do not need to be stored.

The overlap approach hides the infinite sea of heavy fermion states of the extra dimension in the domain wall approach. This tends to obscure the possible presence of singularities in the required inversion of the Wilson kernel. Detailed analysis [174, 175] shows that this operator is particularly well-behaved order by order in perturbation theory. This has led to hopes that this may eventually give a rigorous formulation of chiral models, such as the Standard Model.

Despite being the most elegant known way to have an exact remnant of chiral symmetry on the lattice, the overlap operator raises several issues. These complications probably decrease as the continuum limit is approached, but should be kept in mind given the high computational cost of this approach. To begin with, the overlap is highly non-unique. It explicitly depends on the kernel being projected onto the unitary circle. Even after choosing the Wilson kernel, there is a dependence on the input mass parameter. One might want to define topology in terms of the number of exact zero modes of the overlap operator. However, the non-uniqueness leaves open the question of whether the winding number of a gauge configuration might depend on this choice.

It is important to realize that it is possible to make a bad choice for the hopping parameter for the input kernel. In particular, if it is chosen below the continuum kappa critical value of 1/8, no low modes will survive the construction. This is true despite the fact that the corresponding operator still satisfies the Ginsparg–Wilson condition. This explicitly shows that just satisfying this relation is not sufficient to force a massless fermion mode. Conversely, if one chooses the mass parameter too far in the super-critical region, additional low modes will be produced from the doublers. The Ginsparg–Wilson condition does not immediately determine the number of flavors in the theory.

Another issue concerns the one-flavor case. Because of the anomaly, this theory is not supposed to show any chiral symmetry and has no Goldstone bosons. Nevertheless, one can construct the overlap operator and it will satisfy the Ginsparg–Wilson condition. This shows that the consequences of this condition are weaker than for the usual continuum chiral symmetry. This contrasts sharply with conventional chiral symmetry, where either we have the Goldstone bosons of spontaneous chiral breaking or we have massless fermions [176].

It should also be noted that the overlap operator behaves peculiarly for fermions in higher representations than the fundamental. As previously discussed, the number of zero modes associated with a non-trivial topology in the continuum theory depends on the fermion representation being considered. It has been observed in numerical simulations that the appropriate multiplicity is not always seen for the overlap operator, particularly when constructed on rough gauge configurations [177].

As a final comment, note that these actions preserving a chiral symmetry all involve some amount of non-locality. With minimal doubling, this has a finite range, but is crucial for the anomaly to be allowed to work out properly. An important consequence is that the operator product expansion, a standard perturbative tool, must involve operators with a similar non-locality. The ambiguities in defining non-degenerate quark masses lie in these details.

17.4. Staggered fermions

Another fermion formulation that has an exact chiral symmetry is the so-called "staggered" approach. To derive this, it is convenient to begin with the "naive" discretization of the Dirac equation from before. This considers fermions hopping between nearest neighbor lattice sites while picking up a factor of $\pm i\gamma_\mu$ for a hop in direction $\pm\mu$. Going to momentum space, the discretization replaces powers of momentum with trigonometric functions, for example

$$\gamma_\mu p_\mu \to \gamma_\mu \sin(p_\mu). \tag{17.21}$$

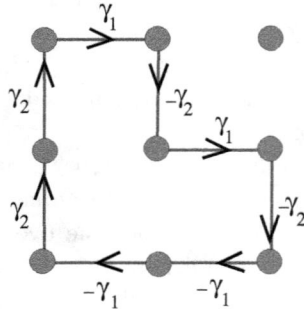

Figure 17.2:   When a fermion circumnavigates a loop in the naive formulation, it picks up a factor that always involves an even power of any particular gamma matrix. This allows the fermion operator to be block-diagonalized into four independent pieces. (Figure from Ref. [147].)

Here, we work in lattice units and thus drop factors of a. As discussed before, this formulation reveals the famous "doubling" issue, arising because the fermion propagator has poles not only for small momentum, but also whenever any component of the momentum is near π. The theory represents not one fermion, but sixteen. And the various doublers have differing chiral properties. This arises from the simple observation that

$$\frac{d}{dp}\sin(p)|_{p=\pi} = -\frac{d}{dp}\sin(p)|_{p=0}. \qquad (17.22)$$

The consequence is that the helicity projectors $(1 \pm \gamma_5)/2$ for a traveling particle depend on which doubler one is observing.

Now consider a fermion traversing a closed loop on the lattice. As illustrated in Fig. 17.2, the corresponding gamma matrix factors will always involve an even number of factors for any particular γ_μ. Thus the resulting product is always proportional to the identity. If a fermion starts off with a particular spinor component, it will wind up in the same component after circumnavigating the loop. This means that the fermion determinant exactly factorizes into four equivalent pieces. The naive theory has an exact $U(4)$ symmetry, as pointed out some time ago by Karsten and Smit [178]. Indeed, for massless fermions this is actually a $U(4) \otimes U(4)$ chiral symmetry. This symmetry does not contradict any anomalies since it is not the full naive $U(16) \otimes U(16)$ of 16 species. The chiral symmetry generated by γ_5 remains exact, but is allowed because it is actually a flavored symmetry. As mentioned above, the helicity projectors for the various doubler species use different signs for γ_5.

The basic idea of staggered fermions is to divide out this $U(4)$ symmetry [179–181] by projecting out a single component of the fermion spinor on each site. Taking $\psi \to P\psi$, one projector that accomplishes this is

$$P = \frac{1}{4}\left(1 + i\gamma_1\gamma_2(-1)^{x_1+x_2} + i\gamma_3\gamma_4(-1)^{x_3+x_4} + \gamma_5(-1)^{x_1+x_2+x_3+x_4}\right),$$

(17.23)

where the x_i are the integer coordinates of the respective lattice sites. This immediately reduces the doubling from a factor of sixteen to four. It is the various oscillating sign factors in this formula that give staggered fermions their name. Since we only need to keep track of one component of ψ at each site, this greatly decreases the computational needs for simulations. This is, to a large extent, responsible for the popularity of this approach.

At this stage, the naive $U(1)$ axial symmetry remains. Indeed, the projector used above commutes with γ_5. This symmetry is allowed since four species, often called "tastes," remain. Among them, the symmetry is a taste non-singlet; under a chiral rotation, two of the doublers rotate one way and two the other.

The next step taken by most of the groups using staggered fermions is the rooting trick. In the hope of reducing the multiplicity down from four, the determinant is replaced with its fourth root, $|D| \to |D|^{1/4}$. With several physical flavors, this trick is applied separately to each. In simple perturbation theory, this is correct since each fermion loop gets multiplied by one quarter, canceling the extra factor from the four "tastes."

At this point one should be extremely uneasy: the exact chiral symmetry is waving a huge red flag. Symmetries of the determinant survive rooting, and thus the exact $U(1)$ axial symmetry for the massless theory remains. For the unrooted theory, this was a flavored chiral symmetry. But, having reduced the theory to one flavor, how can there be a flavored symmetry without multiple flavors? We will now show that this rooting trick can fail non-perturbatively when applied to the staggered quark operator.

In previous chapters, we have seen that the chiral symmetry with N_f fermion flavors has a rather complicated dependence on N_f. With only one flavor there is no chiral symmetry at all, while in general, if the fermions are massless, there are $N_f^2 - 1$ Goldstone bosons. We have also seen a qualitative difference in the mass dependence between an even and an odd number of species. Physics does not behave smoothly in the number of flavors and this raises issues for fermion formulations that inherently have multiple flavors, such as staggered fermions.

Starting with four flavors, the basic question is, can one adjust N_f down to one using the formal expression

$$
\begin{vmatrix}
D+m & 0 & 0 & 0 \\
0 & D+m & 0 & 0 \\
0 & 0 & D+m & 0 \\
0 & 0 & 0 & D+m
\end{vmatrix}^{\frac{1}{4}} = |D+m|? \qquad (17.24)
$$

This has been proposed and is widely used as a method for eliminating the extra species appearing with staggered fermion simulations.

At this point it is important to emphasize that asking about the viability of Eq. (17.24) is a vacuous question outside the context of a regulator. Field theory has divergences that need to be controlled, and, as we have seen above, the appearance of anomalies requires care. In particular, the regulated theory must explicitly break all anomalous symmetries in a way that survives in the continuum limit.

So we must apply Eq. (17.24) before removing the regulator. This is generally expected to be okay as long as the regulator breaks any anomalous symmetries appropriately on each of the four factors. For example, rooting four copies of an overlap operator should be valid as long as the quark mass is positive. Similarly, rooting four equivalent replicas of a massive Wilson fermion operator would be a trivially valid way to obtain the one flavor theory.[2]

The problem is more subtle for attempts to reduce staggered fermions from their inherent four species. In this case the distinct "tastes" are inequivalent and come in chiral pairs. The primary issue is the exact chiral symmetry, which survives the rooting process. This is a higher symmetry than allowed by the flavor singlet chiral anomaly.

Considering several "flavors" of staggered quarks, each with its inherent four "tastes," there will be an exact chiral symmetry associated with each flavor. This symmetry will force the existence of one pseudo-Goldstone boson separately for each flavor. In the three-flavor case there will be one each for the three combinations $i\bar{u}\gamma_5 u$, $i\bar{d}\gamma_5 d$, and $i\bar{s}\gamma_5 s$. Each of these will have an expected squared mass proportional to the corresponding constituent quark. This gives three neutral pseudo-scalars rather than just the expected neutral pion and the eta mesons. The third pseudo-scalar, the eta

---

[2]This requires the mass to be large enough that negative eigenvalues of D do not occur. Otherwise the phase of the fourth root would be ambiguous.

prime, has been given a heavy mass by the anomaly, as extensively discussed in earlier chapters. At a more technical level, these three states in the rooted staggered theory are forbidden to mix, since each is an element of the 15 representation of a separate taste symmetry group.

The conclusion is that rooted staggered fermions are not QCD. So, what is expected to go wrong? The unbroken extra chiral symmetry will give rise to extra minima in the effective potential as a function of σ and η'. In particular, for one-flavor QCD, one will get an effective potential with minima along the lines of Fig. 14.1 instead of the desired structure of Fig. 14.3. Forcing the extra minima would most likely drive the η' mass down from its physical value. This shift should be rather large, of order Λ_{QCD}. This is testable, but being dominated by disconnected diagrams, may be rather difficult to verify in practice. In addition, if we vary the quark masses, the extra minima will result in phase transitions occurring whenever any single quark mass passes through zero. The previous discussion of the one-flavor theory and the two-flavor theory with non-degenerate quarks both show that this is unphysical; no structure is expected when only a single quark mass vanishes.

This problem is admittedly subtle. Formula (17.24) seems intuitively obvious and, as mentioned above, does work if the individual factors take care of the possible anomalies, as with four copies of the overlap operator. Also, as mentioned earlier, rooting is correct in perturbation theory; the rooting factor reduces any internal fermion loop by the correct amount.

Despite these problems, several lattice collaborations continue to pursue staggered fermions using the rooting trick [182–184]. The justification is partly because the simulations are inherently faster than with Wilson fermions, and partly because the exact chiral symmetry simplifies operator mixing. The success of a variety of calculations which are not strongly dependent on the anomaly shows the approach, while technically incorrect, is often a good approximation. On the other hand, if one's goal is to test QCD as the theory of the strong interactions or to estimate QCD corrections to Standard Model processes [185], then one must be extremely wary of any discrepancies found using this method.

If the approach is fundamentally flawed, why do previous calculations with this method frequently appear to be fairly accurate? The issues are connected with so-called "disconnected diagrams." These are fundamental to the mixing between the strange and the light quarks inherent in the eta meson. Most previous staggered calculations have concentrated on particles dominated by valence quarks, ones that propagate without such direct mixing. For these, the problems are swept into the sea quarks, those participating in disconnected closed loops, which are known to contribute on the order of ten percent relative to results in the valence [59] or quenched [186] approximation, where the sea is ignored. Furthermore, the sea quark contributions will primarily differ because of incorrect multiplicities in the "pion cloud," which will have an extra pseudo-Goldstone boson. Thus, the final error for physics dominated by valence quarks is expected to be reduced to a few percent. More serious problems are expected when disconnected diagrams are crucial. This should be particularly serious for the physics of the eta and eta-prime mesons as well as for isospin breaking effects.

Further study

- Consider $SU(2)$ lattice gauge theory coupled to a single spinless fermion degree of freedom. Show that at negative gauge coupling β, the theory is equivalent to staggered fermions at positive β. What happens at negative β if the gauge group is $SU(3)$?

Chapter 18

Quantum fluctuations and topology

We have seen how zero modes of the Dirac operator are closely tied to the chiral anomaly. And we have seen that for smooth classical fields, configurations that give zero modes for the continuum Dirac operator do indeed exist. However, when getting into more detail with defining a lattice Dirac operator, we found subtle issues about which operator to use. And way back in Chapter 5 we saw that typical fields in path integrals are non-differentiable. This leads to the question of uniqueness for the winding number of a given gauge configuration. Indeed, is something like the topological susceptibility of the vacuum a true physical observable?

In Ref. [187], a definition of topological charge was constructed using a fermion operator as a regulator for the trace of γ_5, much as in the earlier derivation of the index theorem. The resulting gauge field operator is defined on hyper-cubes.

For any given hyper-cube, consider all 16 directed hyper-diagonals. For each diagonal, sum over all 24 of the four-hop paths from one corner to the opposite along the hyper-cube edges. Then return to the starting corner, again using all possible paths. With 16 diagonals, this gives a total of $24^2 * 16$ Wilson loops. Finally add these together, giving each path a sign corresponding to the parity of the permutation of the initial four hops.

In practice, one need not actually calculate all these loops individually. Accumulating the forward paths into two matrices, one being the sum of inter-corner paths and one being the sum with the sign factor included, the

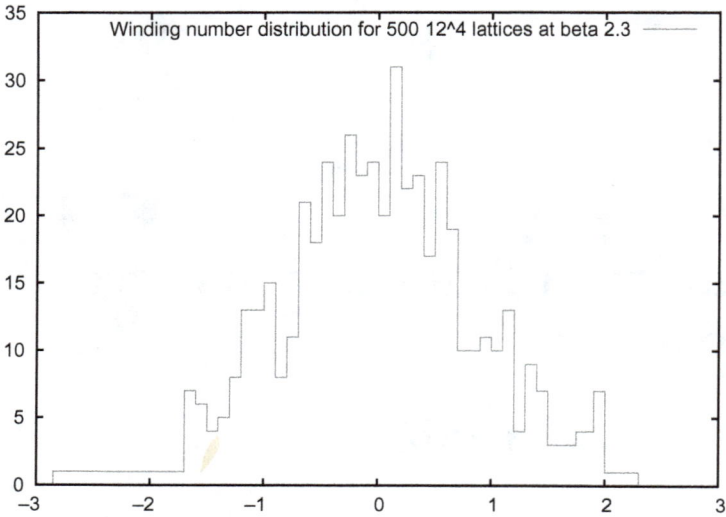

Figure 18.1:   The distribution of the lattice winding number defined in the text on a set of 500 $SU(2)$ lattices of size 12^4 at $\beta = 2.3$. Note the absence of any strong enhancement at integer values. (Figure from Ref. [187]).

desired sum of loops is immediately found from the product of the second sum times the adjoint of the first. Also, each diagonal needs to be calculated only in one direction since the reverse is equivalent.

To normalize this construction, consider small smooth fields and require that this sum of Wilson loops reduce to the classical winding number

$$q \to \nu = \frac{1}{16\pi^2} \int d^4 x \mathrm{Tr}_c F\tilde{F}, \tag{18.1}$$

where Tr_c refers to a trace over color matrices. This algebraic exercise leads to a multiplication of the above sum-over by the factor $\frac{1}{3*2^{10}\pi^2}$. Note that this is independent of the gauge group.

Because typical lattice fields tend to be quite rough, this quantity has no need to peak at integer values. Figure 18.1 shows the distribution of q over a set of 500 independent gauge configurations with gauge group $SU(2)$ at a coupling $\beta = 2.3$ on a 12^4 site lattice.

18.1.   Cooling

It has been known for some time that to expose topological structures in lattice configurations, some cooling procedure to remove short distance

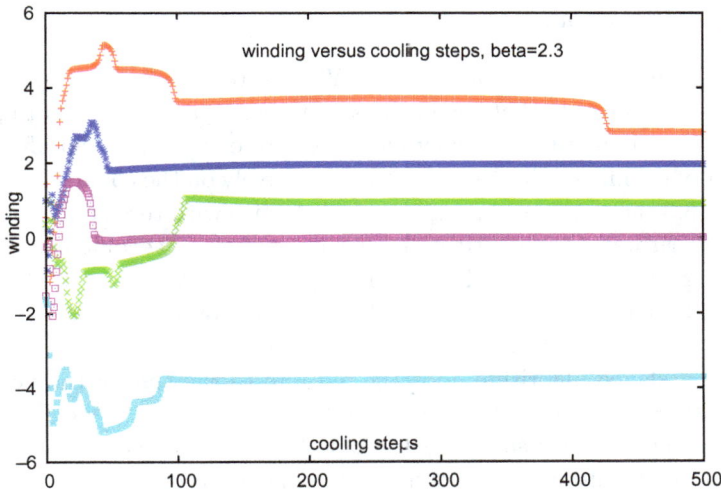

Figure 18.2: The winding number as a function of cooling steps for a set of 5 lattices of size 16^4 for the gauge group $SU(2)$ at $\beta = 2.3$. Note how it settles into approximately integer values with occasional jumps between different windings. (Figure from Ref. [187]).

roughness is required [188]. A particularly simple process is to loop over the lattice in a checkerboard manner and replace each link with a group element that minimizes the total action of the six plaquettes attached to that link. This can be modified by over or under relaxation. Such processes monotonically decrease the total action.

Figure 18.2 shows the evolution of the above winding number under cooling for 5 independent 16^4 site lattices at $\beta = 2.3$. Note how the winding tends to fall into discrete values, with jumps between them as the cooling proceeds further. Empirically, with enough cooling, any $SU(2)$ gauge configuration eventually decays to a state of zero action, gauge equivalent to the vacuum. Indeed, this might be expected since the gauge field space in lattice gauge theory is simply connected.

Since configurations appear to cool ultimately to trivial winding, using a cooling algorithm to define topology requires an arbitrary selection for cooling time. Modifying the Wilson action can prevent the winding decay. For example, constraining the lattice action from any given plaquette from becoming larger than a known small number will prevent instanton decay [189]. Such an "admissibility" condition, however, violates reflection positivity [116] and arbitrarily selects a special minimum instanton size.

18.2.   Uniqueness

Cooling time is not the only issue. While attaining an integer winding requires cooling, note that the initial stages seem quite chaotic in Fig. 18.2. This raises the question of whether the discrete levels reached after some amount of cooling might depend rather sensitively on the cooling algorithm. Figure 18.3 shows the evolution of a single lattice with three different relaxation algorithms. One algorithm is that from above, sweeping over the lattice using checkerboard ordering and replacing each link with the group element that minimizes the action. This is done by projecting the sum of staples that interact with the link onto the group. For the second approach, an under-relaxed algorithm adds the old link to the sum of the neighborhood staples before the projection onto the new group element. Finally, an over-relaxed approach subtracts the old element from the staple sum. We see that the resulting windings not only depend on cooling time, but also on the specific algorithm chosen.

In an extensive analysis, Ref. [190] has compared a variety of filtering methods to expose topological structures in gauge configurations. All schemes have some ambiguities, but on configurations where the

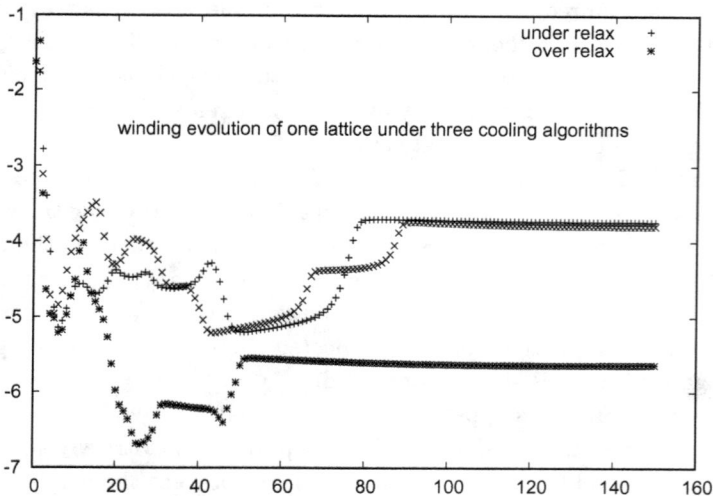

Figure 18.3:   The topological charge evolution for three different cooling algorithms on a single $\beta = 2.3$ lattice configuration for $SU(2)$ on a 16^4 lattice. (Figure from Ref. [187]). With the higher winding numbers, lattice artifacts shift the plateaus slightly away from integer values.

topological structures are clear, the various approaches give similar results. Nevertheless the question remains of whether there is a rigorous and unambiguous definition of topology that applies to typical configurations arising in a simulation. Luscher discussed using a differential flow with the Wilson gauge action to accomplish the cooling [191]. This corresponds to the limit of maximal under-relaxation. This approach still allows topology collapse unless prevented by something like the admissibility condition or the selection of an arbitrary flow time. In addition, if one wishes to determine the topological charge of a configuration obtained in some large scale dynamical simulation, it is unclear why one should take the particular choice of the Wilson gauge action for the cooling procedure.

The sensitivity to the cooling algorithm on rough gauge configurations suggests that there may be an inherent ambiguity in defining the topological charge of typical gauge configurations and consequently, a small ambiguity in the definition of topological susceptibility. It also raises the question of how smooth a given definition of topological charge is as the gauge fields vary. How much correlation is there between the topological number for gauge configurations that are, in some sense, near to each other? Although such issues are quite old [188], they continue to be of considerable interest [192–194].

Topological charge is suppressed when light dynamical quarks are present. This is connected to the question discussed earlier of whether the concept of a single massless quark is well-defined [115]. The chiral limit with multiple massless quarks should give zero topological susceptibility when a chiral fermion operator, such as the overlap, is used. However, with only a single light quark, the lack of chiral symmetry indicates that there is no physical singularity in the continuum theory as this mass passes through zero. Any scheme dependent ambiguity in defining the quark mass would then carry through to the topological susceptibility.

One might argue that the overlap operator solves this problem by defining the winding number as the number of its zero eigenvalues. Indeed, it has been shown [194, 195] that this definition gives a finite result for the susceptibility in the continuum limit. As one is using the fermion operator only as a probe of the gluon fields, this definition can be reformulated directly in terms of the underlying Wilson operator [196]. While the result may have a finite continuum limit, earlier discussion showed that the overlap operator is not unique. In particular, it depends on the initial Dirac operator being projected onto the overlap circle. For the conventional Wilson kernel, there remains a dependence on the input hopping parameter, often referred to as the "domain-wall height." Whether this leaves an ambiguity in the index depends on the density of real eigenvalues of the kernel in the vicinity of where the projection is taken. Numerical evidence [170] suggests that this density decreases with lattice spacing, but it is unclear if this decrease is rapid enough to give a unique susceptibility in the continuum limit. The admissibility condition does eliminate this ambiguity; however, as mentioned earlier, this condition is inconsistent with reflection positivity.

Whether topological susceptibility is well-defined or not seems to have no particular phenomenological consequences. This is not a quantity directly measured in any scattering experiment. It is only defined in the context of a technical definition in a particular non-perturbative simulation. Different valid schemes for regulating the theory might well come up with different values; it is only physical quantities such as hadronic masses that must match between approaches. The famous Witten-Veneziano relation [197, 198] does connect topological susceptibility of the pure gauge theory with the eta prime mass in the limit of large number of colors. This mass, of course, remains well-defined in the physical case of three colors, but the finite N_c corrections to topology can depend delicately on gauge field fluctuations, which are the concern here.

Chapter 19

The Standard Model

So far, we have concentrated on the strong interactions. It is only for this sector of the Standard Model that perturbation theory fails so spectacularly. But the weak and electromagnetic interactions are crucial parts of the full Standard Model.[1] For these interactions, it is also true that the perturbative expansion is not a convergent series. Because the underlying couplings are so small, this does not appear to be of any practical importance.

Nevertheless, it would be conceptually desirable to have a non-perturbative lattice formulation for these interactions as well. From a purist point of view, the continuum limit of a lattice theory provides a precise definition for continuum field theory. Without such for the other interactions, can we really say that the Standard Model is a well-founded quantum field theory? Indeed, it is the problem of chiral gauge theories that encourages studies of chiral symmetry from all possible angles and provides much of the motivation for this book.

In this context we note that the general picture of the Standard Model has changed dramatically over the years. Originally, the successes of quantum electrodynamics made it the model relativistic quantum field theory. Before QCD, the strong interactions were a mystery. But now we see that because of asymptotic freedom, QCD on its own is likely to be a well-defined and self-contained theory. It is the electroweak theory, where both the electric charge and the Higgs couplings are not asymptotically free, for which

---

[1] Gravitation is ignored here because of even more serious unsolved problems.

we lack a non-perturbative formulation. Indeed, a speculative topic such as the possibility of emergent gravity may be intimately tied to these issues with the weaker forces.

For the electromagnetic interactions, a lattice formulation at first seems straightforward, involving the introduction of a $U(1)$ gauge field for the photons. Unlike the strong interaction case, however, we do not have asymptotic freedom to tell us how to take the continuum limit. And the physical coupling $\alpha \sim 1/137$ seems to be an unnaturally small number. Perhaps electrodynamics on its own does not actually exist as a field theory, as believed to be the case for the scalar ϕ^4 theory. But photons and electrons are essential components to the world around us. One interesting possibility is that electromagnetism is actually only a part of a higher level theory, perhaps in some unification with the strong interactions.

We hit a more serious snag with the weak interactions; the couplings violate parity. The W bosons appear to interact only with left handed fermions. We need to couple the fermions in a chiral manner. It is not known how to do this in any non-perturbative scheme. The problem here is closely tied to the chiral anomaly and the fact that not all currents can be simultaneously conserved. When applied to the weak interactions, the 't Hooft vertex gives rise to effective interactions that do not conserve baryon number. Any complete non-perturbative formulation must allow for such processes [199].

Some attempts to include the Standard Model in a domain wall formulation have been presented [155, 200], but these generally involve heavy additional states such as "mirror" fermions [201]. While possibly viable, such approaches lack the theoretical elegance of the original Wilson lattice gauge theory. Also, it is not clear how to avoid having the mirror states affect Higgs phenomenology through virtual loops.

Perhaps a lattice formulation more intimately tied to unification ideas could help here. The group $SO(10)$ looks quite interesting in this context [108]. Here, a single generation of fermions fits nicely into a single 16 dimensional representation of this group. And, anomalies are automatically canceled in this picture. This would seem to indicate that there should be no obvious requirement for doublers as an obstacle to a lattice construction. However, the usual Wilson approach seems to require a term that is not a singlet under this group. This could be overcome with some added Higgs-like scalar fields, but then we get closer to the models mentioned above, with the doublers playing the role of mirror fermions.

19.1. Where is the parity violation?

The Standard Model of elementary particle interactions is based on the product of three gauge groups, $SU(3) \otimes SU(2) \otimes U(1)_{em}$. Here, the $SU(3)$ represents the strong interactions of quarks and gluons, the $U(1)_{em}$ corresponds to electromagnetism, and the $SU(2)$ gives rise to the weak interactions. For simplicity, we ignore the technical details of electroweak mixing here.

The full Model is, of course, parity-violating, as necessary to describe observed helicities in beta decay. This violation is normally considered to lie in the $SU(2)$ of the weak interactions, with both the $SU(3)$ and $U(1)_{em}$ being parity-conserving. However, this is actually only a convention, adopted primarily because the weak interactions are small compared to the others. It is possible to reinterpret the various degrees of freedom such that the $SU(2)$ gauge interaction is vector-like. Since the full Model breaks parity, this process shifts the violation out of the weak and into the strong, electromagnetic, and Higgs interactions. The resulting theory pairs the left handed electron with a right handed anti-quark to form a Dirac fermion. With a vector-like weak interaction, the chiral issues which complicate lattice formulations now move to the other gauge groups. Requiring gauge invariance for the re-expressed electromagnetism then clarifies the mechanism behind one speculative proposal for a lattice regularization of the standard model [88, 202].

For simplicity, consider only the first generation, which involves four left handed doublets. These correspond to the neutrino/electron lepton pair plus three colors for the up/down quarks

$$\begin{pmatrix} \nu \\ e^- \end{pmatrix}_L, \begin{pmatrix} u^r \\ d^r \end{pmatrix}_L, \begin{pmatrix} u^g \\ d^g \end{pmatrix}_L, \begin{pmatrix} u^b \\ d^b \end{pmatrix}_L. \tag{19.1}$$

The superscripts from the set $\{r, g, b\}$ represent the internal $SU(3)$ index of the strong interactions, and the subscript L indicates left handed helicities.

If we ignore the strong and electromagnetic interactions, leaving only the weak $SU(2)$, each of these four doublets is equivalent and independent. We now arbitrarily pick two of them and do a charge conjugation operation, thus switching to their antiparticles:

$$\begin{pmatrix} u^g \\ d^g \end{pmatrix}_L \longrightarrow \begin{pmatrix} \overline{d^g} \\ \overline{u^g} \end{pmatrix}_R,$$

$$\begin{pmatrix} u^b \\ d^b \end{pmatrix}_L \longrightarrow \begin{pmatrix} \overline{d^b} \\ \overline{u^b} \end{pmatrix}_R. \tag{19.2}$$

In four dimensions, anti-fermions have the opposite helicity; so, we label these new doublets with R representing right handedness.

With two left and two right handed doublets, we can combine them into two Dirac doublets

$$
\left(\begin{array}{c} \begin{pmatrix} \nu \\ e^- \end{pmatrix}_L \\ \begin{pmatrix} \overline{d^g} \\ \overline{u^g} \end{pmatrix}_R \end{array}\right)
\qquad
\left(\begin{array}{c} \begin{pmatrix} u^r \\ d^r \end{pmatrix}_L \\ \begin{pmatrix} \overline{d^b} \\ \overline{u^b} \end{pmatrix}_R \end{array}\right).
\tag{19.3}
$$

Formally in terms of the underlying fields, the construction takes

$$
\psi = \frac{1}{2}(1 - \gamma_5)\psi_{(\nu,e^-)} + \frac{1}{2}(1 + \gamma_5)\psi_{(\overline{d^g},\overline{u^g})},
\tag{19.4}
$$

$$
\chi = \frac{1}{2}(1 - \gamma_5)\psi_{(u^r,d^r)} + \frac{1}{2}(1 + \gamma_5)\psi_{(\overline{d^b},\overline{u^b})}.
\tag{19.5}
$$

From the conventional point of view, these fields have rather peculiar quantum numbers. For example, the left and right parts have different electric charges. Electromagnetism now violates parity. The left and right parts also have different strong quantum numbers; the strong interactions violate parity as well. Finally, the components have different masses; parity is violated in the Higgs mechanism.

The different helicities of these fields also have variant baryon number. This is directly related to the known baryon violating processes through weak "instantons" and axial anomalies [26]. When a topologically non-trivial weak field is present, the axial anomaly arises from a level flow out of the Dirac sea [78]. This generates a spin flip in the fields, *i.e.* $e_L^- \to (\overline{u^g})_R$. Because of the peculiar particle identification above, this process does not conserve charge, with $\Delta Q = -\frac{2}{3} + 1 = \frac{1}{3}$. This would be a disaster for electromagnetism were it not for the fact that simultaneously, the other Dirac doublet also flips $d^r{}_L \to (\overline{u^b})_R$ with a compensating $\Delta Q = -\frac{1}{3}$. This is anomaly cancellation, with the total $\Delta Q = \frac{1}{3} - \frac{1}{3} = 0$. Only when both doublets are considered together is the $U(1)$ symmetry restored. In this process, baryon number is violated, with $L + Q \to \overline{Q} + \overline{Q}$. This is a consequence of the famous "'t Hooft vertex" [26] discussed earlier in the context of the strong interactions.

19.2.   A lattice model

The above discussion on twisting the gauge groups is for the continuum picture. Now we turn to the lattice and show how this picture leads to a possible lattice model for the strong interactions, albeit with an unusual added coupling that renders the treatment quite difficult to make rigorous [88, 202]. Whether this model is viable remains undecided, but it does incorporate many of the required features.

The extra coupling in this approach is formally an irrelevant operator in the sense discussed in Chapter 9. Thus, naively, it should not affect the continuum limit. However, as we have seen with the added term for Wilson fermions, a formally irrelevant term could drive one to a different renormalization group fixed point. This term drastically modifies the number of surviving fermion fields. In the present case we might hope for a parallel modification of the continuum theory, one in which only the proper spectrum and chiral couplings survive.

We base the following discussion in the language of the domain wall approach for the fermions [135, 156]. Our four-dimensional world is modeled as a "4-brane" embedded in five dimensions. The complete lattice is a five-dimensional box with open boundaries. The parameters are chosen so that the physical quarks and leptons appear as surface zero modes. The elegance of this scheme lies in the natural chirality of these modes as the size of the extra dimension grows. With a finite fifth dimension, one doubling remains, coming from interfaces appearing as surface/anti-surface pairs. It is natural to couple a four-dimensional gauge field equally to both surfaces, giving rise to a vector-like theory.

We now insert the above pairing into this five-dimensional scheme. In particular, consider the left handed electron as a zero mode on one wall and the right handed anti-green-up-quark as the partner zero mode on the other wall, as sketched in Fig. 19.1. This provides a lattice regularization for the $SU(2)$ of the weak interactions.

However, since these two particles have a different electric charge, $U(1)_{EM}$ must be broken somewhere. We consider this breaking to take place in the interior of the extra dimension. Proceeding in analogy to the

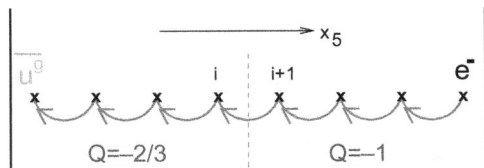

Figure 19.1: Pairing the electron with the anti-green-up-quark. (Figure taken from Ref. [202].)

"wave guide" picture [203], restrict this charge violation to one layer at some interior position $x_5 = i$. Using Wilson fermions, the hopping term from $x_5 = i$ to $i + 1$

$$\overline{\psi}_i P \psi_{i+1} \qquad (P = (\gamma_5 + r)/2) \qquad (19.6)$$

is a $Q = 1/3$ operator. At this layer, electric charge is not conserved. This is unacceptable and needs to be fixed.

To restore the $U(1)$ symmetry, one must transfer the charge from ψ to the compensating doublet χ. For this we replace the sum of hoppings with a product on the offending layer:

$$\overline{\psi}_i P \psi_{i+1} + \overline{\chi}_i P \chi_{i+1} \longrightarrow \overline{\psi}_i P \psi_{i+1} \times \overline{\chi}_i P \chi_{i+1}. \qquad (19.7)$$

This introduces an electrically neutral four-fermion operator. Note that it is baryon violating, involving a "lepto-quark/diquark" exchange, as sketched in Fig. 19.2. In some sense, this operator represents a "filter" at $x_5 = i$, through which only charge compensating pairs of fermions can pass. There is no chiral symmetry in five dimensions. Even for the free theory, combinations like $\overline{\psi}_i P \psi_{i+1}$ have non-vanishing vacuum expectation values. They might be thought of as generating a "tadpole," with χ generating an effective hopping for ψ and vice-versa.

Actually, the above four-fermion operator is not quite sufficient for all chiral anomalies, which can also involve right handed singlet fermions. To correct this, we need to explicitly include the right-hand sector, adding similar four-fermion couplings (also electrically neutral). The main difference is that this sector does not couple to the weak bosons.

Having fixed the $U(1)$ of electromagnetism, we restore the strong $SU(3)$ with an anti-symmetrization of the quark color indices in the new operator, $Q^r Q^g Q^b \longrightarrow \epsilon^{\alpha\beta\gamma} Q^\alpha Q^\beta Q^\gamma$. Note that similar left-right

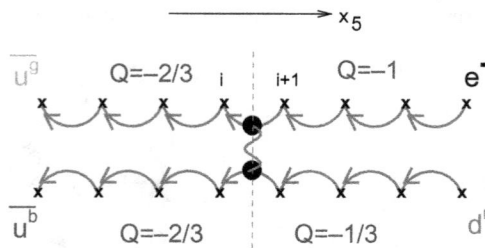

Figure 19.2:   Transferring charge between the doublets. (Figure taken from Ref. [202]).

inter-sector couplings are needed to correctly obtain the effects of topologically non-trivial strong gauge fields.

An alternative view of this picture folds the lattice about the interior of the fifth dimension, placing all light modes on one wall and having the multi-fermion operator on the other. This is the model of Ref. [88], with the additional inter-sector couplings correcting a technical error [204].

Unfortunately, the scheme is still not completely rigorously formulated. In particular, the non-trivial four-fermion coupling represents a new defect and we need to show that this does not give rise to unwanted extra zero modes. Note, however, that the five-dimensional mass is the same on both sides of the defect; thus, there are no topological reasons for such to appear.

A second worry is that the four-fermion coupling might induce an unwanted spontaneous symmetry breaking of one of the gauge symmetries. We need to remain in a strongly coupled paramagnetic phase without this. Ref. [88] showed that strongly coupled zero modes do preserve the desired symmetries, but the analysis ignored contributions from heavy modes in the fifth dimension.

Assuming all works as desired, the model raises several other interesting questions. As formulated, we need a right-handed neutrino to provide all quark states with partners. Is there some variation that avoids this particle, which decouples from all gauge fields in the continuum limit? Another question concerns possible numerical simulations; is the effective action positive? Finally, we have used the details of the usual Standard Model, leaving open the question of whether this model is somehow special. Can we always find an appropriate multi-fermion coupling to eliminate undesired modes in other chiral theories where anomalies are properly canceled?

Further study

- The electroweak theory involves the non-Abelian group $SU(2)$. But gauge theories with non-Abelian groups are supposed to confine. Does this $SU(2)$ confine? What about the fact that electrons appear free? For an interesting take on this issue, see Fradkin and Shenker [205].

Chapter 20

Final remarks

We have explored how non-perturbative physics is required in order to understand many features of QCD. This is particularly important for the phenomena of confinement and chiral symmetry breaking. Issues such as doubling and anomalies suggest at first that these two features do not mesh well. However, in this book we have argued that these actually combine into an elegant and coherent picture. Of particular interest is how chiral symmetry is broken in three rather different ways. The interplay of these mechanisms is the fascinating physics crucial to the final picture.

The first and perhaps most important effect is the dynamical symmetry breaking that leads to the pions being light pseudo-Goldstone bosons. Their dynamics represents the dominant physics for QCD at low energies. The popular and useful expansion of chiral perturbation theory is a natural expansion in the momenta and masses of these particles.

Second, we have the anomaly. This is responsible for eliminating the flavor-singlet axial $U(1)$ symmetry of the classical theory. Thus, the η' meson is not a Goldstone boson and acquires a mass of order Λ_{qcd}. We have seen how understanding this breaking involves non-perturbative physics associated with the zero modes of the Dirac operator and the connection with topological features in the underlying gauge fields.

The third is the explicit symmetry breaking from the quark masses. This is responsible for the pseudo-scalar mesons not being exactly massless. Using the freedom to redefine fields using chiral rotations, the number of independent mass parameters is $N_f + 1$ where N_f is the number of fermion species under consideration. The plus one is a crucial feature allowing for the possibility of CP violation coming from the interplay of the mass term with the anomaly.

Throughout, we have used only a few widely accepted assumptions, such as the existence of QCD as a field theory and standard ideas about chiral symmetry. Since the dominant understanding of quantum field theory comes from perturbation theory, perhaps it is not surprising that some aspects covered here contradict common lore. One is that chiral symmetry is completely lost in a theory with only one light quark. This introduces a non-perturbative additive renormalization of the quark mass. This feature precludes the use of a massless up quark to solve the strong CP problem. Closely tied with this is an inherent ambiguity in defining topological susceptibility. This is non-trivial since non-differentiable fields are known to dominate the path integral.

As simple as the overall picture is, we require effects that go well beyond perturbation theory. Some non-perturbative aspects appear already in the classical theory, although their true importance only appears in the context of the anomaly. We need aspects of the Dirac spectrum when the gauge fields have non-trivial topology, but both have inherent ambiguities. Since the lattice is our only true non-perturbative regulator, the inclusion of this physics properly in such a formulation remains a rich topic for active research.

Bibliography

[1] M. Creutz. Confinement, chiral symmetry, and the lattice. *Acta Physica Slovaca*, 61:1–127, 2011.

[2] M. Gell-Mann. A schematic model of baryons and mesons. *Phys. Lett.*, 8:214–215, 1964.

[3] G. Zweig. An SU(3) model for strong interaction symmetry and its breaking. Version 2. In D.B. Lichtenberg and Simon Peter Rosen, editors, *Developments in the Quark Theory of Hadrons. Vol. 1. 1964–1978*, pages 22–101. 1964.

[4] S. R. Mishra and F. Sciulli. Deep inelastic lepton — nucleon scattering. *Ann. Rev. Nucl. Part. Sci.*, 39:259–310, 1989.

[5] P. D. B. Collins. *An introduction to Regge theory and high-energy physics.* Cambridge, 1977.

[6] E. Eichten, K. Gottfried, T. Kinoshita, K. D. Lane, and T.-M. Yan. Charmonium: Comparison with experiment. *Phys. Rev.*, D21:203, 1980.

[7] W. Meissner and R. Ochsenfeld. Ein neuer effekt bei eintritt der supraleitfhigkeit. *Naturwissenschaften*, 21:787–788, 1933.

[8] A. A. Abrikosov. On the magnetic properties of superconductors of the second group. *Sov. Phys. JETP*, 5:1174–1182, 1957.

[9] H. F. Hess, R. B. Robinson, R. C. Dynes, J. M. Valles, and J. V. Waszczak. Scanning-tunneling-microscope observation of the Abrikosov flux lattice and the density of states near and inside a fluxoid. *Phys. Rev. Lett.*, 62:214–216, 1989.

[10] A. Chodos, R. L. Jaffe, K. Johnson, C. B. Thorn, and V. F. Weisskopf. A new extended model of hadrons. *Phys. Rev.*, D9:3471–3495, 1974.

[11] C.-N. Yang and R. L. Mills. Conservation of Isotopic Spin and Isotopic Gauge Invariance. *Phys. Rev.*, 96:191–195, 1954.

[12] K. G. Wilson. Confinement of quarks. *Phys. Rev.*, D10:2445–2459, 1974.

[13] K. G. Wilson. Quarks and strings on a lattice. *Erice Lectures 1975*, 1977. New phenomena in subnuclear physics. Part A. Proceedings of the First

Half of the 1975 International School of Subnuclear Physics, Erice, Sicily, July 11–August 1, 1975, ed. A. Zichichi, Plenum Press, New York, 1977, p. 69, CLNS-321.

[14] F. J. Dyson. Divergence of perturbation theory in quantum electrodynamics. *Phys. Rev.*, 85:631–632, 1952.

[15] J. Frohlich. On the triviality of lambda (phi**4) in D-dimensions theories and the approach to the critical point in D ≥ four-dimensions. *Nucl. Phys.*, B200:281–296, 1982.

[16] L. D. Landau and I. Ya. Pomeranchuk. On point interactions in quantum electrodynamics. *Dokl.Akad.Nauk Ser.Fiz.*, 102:489, 1955. Also published in Collected Papers of L.D. Landua. Edited by D. Ter Haar. Pergamon Press, 1965. pp. 654–658.

[17] S. Mandelstam. Soliton operators for the quantized Sine-Gordon equation. *Phys. Rev.*, D11:3026, 1975.

[18] S. R. Coleman. Classical lumps and their quantum descendents. *Subnucl. Ser.*, 13:297, 1977.

[19] J. S. Schwinger. Gauge invariance and mass. 2. *Phys. Rev.*, 128:2425–2429, 1962.

[20] S. R. Coleman. The quantum Sine-Gordon equation as the massive Thirring Model. *Phys. Rev.*, D11:2088, 1975.

[21] W. E. Thirring. A soluble relativistic field theory? *Annals Phys.*, 3:91–112, 1958.

[22] C. Itzykson and J. M. Drouffe. *Statistical field theory. Vol. 1: From Brownian motion to renormalization and lattice gauge theory.* Cambridge, 1989.

[23] L. Onsager. Crystal statistics. 1. A Two-dimensional model with an order disorder transition. *Phys. Rev.*, 65:117–149, 1944.

[24] M. Creutz. Hidden symmetries in two dimensional field theory. *Annals Phys.*, 321:2782–2792, 2006.

[25] R. F. Dashen. Some features of chiral symmetry breaking. *Phys. Rev.*, D3:1879–1889, 1971.

[26] G. 't Hooft. Computation of the quantum effects due to a four-dimensional pseudoparticle. *Phys. Rev.*, D14:3432–3450, 1976.

[27] E. Witten. Large N chiral dynamics. *Annals Phys.*, 128:363, 1980.

[28] C. Rosenzweig, J. Schechter, and C. G. Trahern. Is the effective lagrangian for QCD a Sigma model? *Phys. Rev.*, D21:3388, 1980.

[29] R. L. Arnowitt and P. Nath. Effective Lagrangians with U(1) axial anomaly. 1. *Nucl. Phys.*, B209:234, 1982.

[30] P. Nath and R. L. Arnowitt. Effective Lagrangians with U(1) axial anomaly. 2. *Nucl. Phys.*, B209:251, 1982.

[31] K. Kawarabayashi and N. Ohta. The problem of eta in the large N limit: effective Lagrangian approach. *Nucl. Phys.*, B175:477, 1980.

[32] K. Kawarabayashi and N. Ohta. On the partial conservation of the U(1) current. *Prog. Theor. Phys.*, 66:1789, 1981.

[33] N. Ohta. Vacuum structure and chiral charge quantization in the large N limit. *Prog. Theor. Phys.*, 66:1408, 1981.

[34] T. Banks and A. Zaks. On the phase structure of vector-like gauge theories with massless fermions. *Nucl.Phys.*, B196:189, 1982.

[35] S. R. Coleman. There are no Goldstone bosons in two-dimensions. *Commun. Math. Phys.*, 31:259–264, 1973.

[36] E. W. Weisstein. Gompertz constant. http://mathworld.wolfram.com/GompertzConstant.html.

[37] E. Ising. Contribution to the theory of ferromagnetism. *Z. Phys.*, 31:253–258, 1925.

[38] B. M. McCoy and T. T. Wu. *The two-dimensional Ising model.* Harvard University Press, 1973.

[39] M. Creutz. Gauge fixing, the transfer matrix, and confinement on a lattice. *Phys. Rev.*, D15:1128, 1977.

[40] R. P. Feynman. Space-time approach to nonrelativistic quantum mechanics. *Rev. Mod. Phys.*, 20:367–387, 1948.

[41] R. P. Feynman and A. R. Hibbs. *Quantum mechanics and path integrals.* McGraw Hill, 1965.

[42] M. Creutz and B. Freedman. A statistical approach to quantum mechanics. *Ann. Phys.*, 132:427, 1981.

[43] M. Creutz. Evaluating Grassmann integrals. *Phys. Rev. Lett.*, 81:3555–3558, 1998.

[44] M. Creutz. Transfer matrices and lattice fermions at finite density. *Found. Phys.*, 30:487–492, 2000.

[45] S. Weinberg. Feynman Rules for Any Spin. 2. Massless Particles. *Phys. Rev.*, 134:B882–B896, 1964.

[46] M. Creutz. Quarks, gluons, and lattices. 1983. Cambridge, Uk: Univ. Pr. (1983) 169 P. (Cambridge Monographs On Mathematical Physics).

[47] C. Gattringer and C. B. Lang. Quantum chromodynamics on the lattice. *Springer Lect. Notes Phys.*, 788:1–211, 2010.

[48] I. Montvay and G. Munster. *Quantum fields on a lattice.* Cambridge, UK: Univ. Pr., 1994.

[49] T. DeGrand and C. E. Detar. *Lattice methods for quantum chromodynamics.* World Scientific, 2006.

[50] H. J. Rothe. Lattice gauge theories: an introduction. *World Sci. Lect. Notes Phys.*, 74:1–605, 2005.

[51] S. Elitzur. Impossibility of spontaneously breaking local symmetries. *Phys. Rev.*, D12:3978–3982, 1975.

[52] L. D. Faddeev and V. N. Popov. Feynman diagrams for the Yang-Mills field. *Phys. Lett.*, B25:29–30, 1967.

[53] I. M. Singer. Some remarks on the Gribov ambiguity. *Commun. Math. Phys.*, 60:7–12, 1978.

[54] N. Vandersickel and D. Zwanziger. The Gribov problem and QCD dynamics. *Phys. Rept.*, 520:175–251, 2012.

[55] M. Creutz. Monte Carlo study of quantized SU(2) gauge theory. *Phys. Rev.*, D21:2308–2315, 1980.

[56] M. Creutz, L. Jacobs, and C. Rebbi. Monte Carlo study of Abelian lattice gauge theories. *Phys. Rev.*, D20:1915, 1979.

[57] N. Metropolis, A. W. Rosenbluth, M. N. Rosenbluth, A. H. Teller, and E. Teller. Equation of state calculations by fast computing machines. *J. Chem. Phys.*, 21:1087–1092, 1953.

[58] N. Madras and A. D. Sokal. The Pivot algorithm: a highly efficient Monte Carlo method for selfavoiding walk. *J. Statist. Phys.*, 50:109–186, 1988.

[59] D. H. Weingarten and D. N. Petcher. Monte Carlo integration for lattice gauge theories with fermions. *Phys. Lett.*, B99:333, 1981.

[60] F. Fucito, E. Marinari, G. Parisi, and C. Rebbi. A proposal for Monte Carlo simulations of fermionic Systems. *Nucl. Phys.*, B180:369, 1981.

[61] D. J. Scalapino and R. L. Sugar. A method for performing Monte Carlo calculations for systems with fermions. *Phys. Rev. Lett.*, 46:519, 1981.

[62] S. Duane, A. D. Kennedy, B. J. Pendleton, and D. Roweth. Hybrid Monte Carlo. *Phys. Lett.*, B195:216–222, 1987.

[63] M. Creutz. Global Monte Carlo algorithms for many-fermion systems. *Phys. Rev.*, D38:1228–1238, 1988.

[64] H. D. Politzer. Reliable perturbative results for strong interactions? *Phys. Rev. Lett.*, 30:1346–1349, 1973.

[65] D. J. Gross and F. Wilczek. Ultraviolet behavior of nonabelian Gauge theories. *Phys. Rev. Lett.*, 30:1343–1346, 1973.

[66] D. J. Gross and F. Wilczek. Asymptotically free gauge theories. 1. *Phys. Rev.*, D8:3633–3652, 1973.

[67] W. E. Caswell. Asymptotic behavior of nonabelian gauge theories to two loop order. *Phys. Rev. Lett.*, 33:244, 1974.

[68] D. R. T. Jones. Two loop diagrams in Yang-Mills theory. *Nucl. Phys.*, B75:531, 1974.

[69] S. R. Coleman and E. J. Weinberg. Radiative corrections as the origin of spontaneous symmetry breaking. *Phys. Rev.*, D7:1888–1910, 1973.

[70] J. A. M. Vermaseren, S. A. Larin, and T. van Ritbergen. The four loop quark mass anomalous dimension and the invariant quark mass. *Phys. Lett.*, B405:327–333, 1997.

[71] K. G. Chetyrkin, M. Misiak, and M. Munz. Beta functions and anomalous dimensions up to three loops. *Nucl. Phys.*, B518:473–494, 1998.

[72] The Whys of Subnuclear Physics. Proceedings of the 1977 International School of Subnuclear Physics, Held in Erice, Trapani, Sicily, July 23–August 10, 1977, 1979.

[73] A. Hasenfratz and P. Hasenfratz. The connection between the Lambda parameters of lattice and continuum QCD. *Phys. Lett.*, B93:165, 1980.

[74] A. Hasenfratz and P. Hasenfratz. The scales of Euclidean and Hamiltonian lattice QCD. *Nucl. Phys.*, B193:210, 1981.

[75] R. F. Dashen and D. J. Gross. The relationship between lattice and continuum definitions of the gauge theory coupling. *Phys. Rev.*, D23:2340, 1981.

[76] K. G. Wilson and J. B. Kogut. The Renormalization group and the epsilon expansion. *Phys. Rept.*, 12:75–200, 1974.

[77] R. Rajaraman. *Solitons and instantons. An introduction to solitons and instantons in quantum field theory.* North-holland, 1982.

[78] J. Ambjorn, J. Greensite, and C. Peterson. The axial anomaly and the lattice dirac sea. *Nucl. Phys.*, B221:381, 1983.

[79] B. Kelvin and W. Thomson. *Baltimore lectures on molecular dynamics and the wave theory of light*. London : C. J. Clay and sons; Baltimore, Publication agency of the Johns Hopkins university, 1904.

[80] E. Vicari and H. Panagopoulos. Theta dependence of $SU(N)$ gauge theories in the presence of a topological term. *Phys. Rept.*, 470:93–150, 2009.

[81] J. Goldstone, A. Salam, and S. Weinberg. Broken Symmetries. *Phys. Rev.*, 127:965–970, 1962.

[82] J. S. Schwinger. Field theory commutators. *Phys. Rev. Lett.*, 3:296–297, 1959.

[83] M. Creutz. Anomalies and chiral symmetry in QCD. *Annals Phys.*, 324:1573–1584, 2009.

[84] J. D. Bjorken. Inequality for backward electron-nucleon and muon-nucleon scattering at high momentum transfer. *Phys. Rev.*, 163:1767–1769, 1967.

[85] J. D. Bjorken. Applications of the chiral $U(6) \otimes U(6)$ algebra of current densities. *Phys. Rev.*, 148:1467–1478, 1966.

[86] J. Wess and B. Zumino. Consequences of anomalous Ward identities. *Phys. Lett.*, 37B:95–97, 1971.

[87] E. Witten. Global aspects of current algebra. *Nucl. Phys.*, B223:422–432, 1983.

[88] M. Creutz, M. Tytgat, C. Rebbi, and S.-S. Xue. Lattice formulation of the standard model. *Phys. Lett.*, B402:341–345, 1997.

[89] J. Gasser and H. Leutwyler. Quark Masses. *Phys. Rept.*, 87:77–169, 1982.

[90] D. B. Kaplan and A. V. Manohar. Current mass ratios of the light quarks. *Phys. Rev. Lett.*, 56:2004, 1986.

[91] S. Weinberg. The problem of mass. *Trans. New York Acad. Sci.*, 38:185–201, 1977.

[92] H. Leutwyler. The ratios of the light quark masses. *Phys. Lett.*, B378:313–318, 1996.

[93] M. Creutz. Spontaneous violation of CP symmetry in the strong interactions. *Phys. Rev. Lett.*, 92:201601, 2004.

[94] S. L. Adler. Axial vector vertex in spinor electrodynamics. *Phys. Rev.*, 177:2426–2438, 1969.

[95] S. L. Adler and W. A. Bardeen. Absence of higher order corrections in the anomalous axial vector divergence equation. *Phys. Rev.*, 182:1517–1536, 1969.

[96] J. S. Bell and R. Jackiw. A PCAC puzzle: pi0 to gamma gamma in the sigma model. *Nuovo Cim.*, A60:47–61, 1969.

[97] K. Fujikawa. Path integral measure for gauge invariant fermion theories. *Phys. Rev. Lett.*, 42:1195, 1979.

[98] S. R. Coleman. The uses of instantons. *Subnucl. Ser.*, 15:805, 1979.

[99] P. H. Ginsparg and K. G. Wilson. A remnant of chiral symmetry on the lattice. *Phys. Rev.*, D25:2649, 1982.

[100] H. Neuberger. Exactly massless quarks on the lattice. *Phys. Lett.*, B417:141–144, 1998.

[101] M. Creutz. Quark masses and chiral symmetry. *Phys. Rev.*, D52:2951–2959, 1995.

[102] T. Banks and A. Casher. Chiral symmetry breaking in confining theories. *Nucl. Phys.*, B169:103, 1980.

[103] E. Witten. Constraints on supersymmetry breaking. *Nucl. Phys.*, B202:253, 1982.

[104] E. Corrigan and P. Ramond. A note on the quark content of large color groups. *Phys. Lett.*, B87:73, 1979.

[105] A. Armoni, M. Shifman, and G. Veneziano. SUSY relics in one flavor QCD from a new 1/N expansion. *Phys. Rev. Lett.*, 91:191601, 2003.

[106] F. Sannino and M. Shifman. Effective Lagrangians for orientifold theories. *Phys. Rev.*, D69:125004, 2004.

[107] M. Unsal and L. G. Yaffe. (In)validity of large N orientifold equivalence. *Phys. Rev.*, D74:105019, 2006.

[108] H. Georgi and D. V. Nanopoulos. Ordinary predictions from grand principles: T quark mass in O(10). *Nucl. Phys.*, B155:52, 1979.

[109] S. Okubo. Phi meson and unitary symmetry model. *Phys. Lett.*, 5:165–168, 1963.

[110] J. Iizuka. Systematics and phenomenology of meson family. *Prog. Theor. Phys. Suppl.*, 37:21–34, 1966.

[111] C. Vafa and E. Witten. Parity conservation in QCD. *Phys. Rev. Lett.*, 53:535, 1984.

[112] M. Creutz. One flavor QCD. *Annals Phys.*, 322:1518–1540, 2007.

[113] M. Creutz. The 't Hooft vertex revisited. *Annals Phys.*, 323:2349–2365, 2008.

[114] M. Creutz. Comments on staggered fermions / Panel discussion. *PoS*, CONFINEMENT8:016, 2008.

[115] M. Creutz. Ambiguities in the up quark mass. *Phys. Rev. Lett.*, 92:162003, 2004.

[116] M. Creutz. Positivity and topology in lattice gauge theory. *Phys. Rev.*, D70:091501, 2004.

[117] G. 't Hooft. Aspects of quark confinement. *Phys. Scripta*, 24:841–846, 1981.

[118] D. Gaiotto, A. Kapustin, Z. Komargodski, and N. Seiberg. Theta, Time reversal, and temperature. *JHEP*, 05:091, 2017.

[119] M. Creutz. Quark mass dependence of two-flavor QCD. *Phys. Rev.*, D83:016005, 2011.

[120] H. Georgi and I. N. McArthur. Instantons and the m_u quark mass. *unpublished (HUTP-81/A011)*, 1981.

[121] T. Banks, Y. Nir, and N. Seiberg. Missing (up) mass, accidental anomalous symmetries, and the strong CP problem. *unpublished (hep-ph/9403203)*, 1994.

[122] K. Choi, C. W. Kim, and W. K. Sze. Mass renormalization by instantons and the strong CP problem. *Phys. Rev. Lett.*, 61:794, 1988.

[123] R. Frezzotti, P. A. Grassi, S. Sint, and P. Weisz. Lattice QCD with a chirally twisted mass term. *JHEP*, 0108:058, 2001.

[124] Ph. Boucaud *et al.* Dynamical twisted mass fermions with light quarks. *Phys. Lett.*, B650:304–311, 2007.

[125] V. Baluni. CP Violating Effects in QCD. *Phys. Rev.*, D19:2227–2230, 1979.

[126] S. Aoki. New phase structure for lattice QCD with Wilson fermions. *Phys. Rev.*, D30:2653, 1984.

[127] H. B. Nielsen and M. Ninomiya. Absence of neutrinos on a lattice. 1. proof by homotopy theory. *Nucl. Phys.*, B185:20, 1981.

[128] M. Creutz. Effective potentials, thermodynamics, and twisted mass quarks. *Phys. Rev.*, D76:054501, 2007.

[129] M. Creutz. Wilson fermions at finite temperature. 1996. RHIC Summer Study '96: Theory workshop on relativistic heavy ion collisions, D. Kahana and Y. Pang, eds., pp. 49–54 (NTIS, 1997); hep-lat/9608024.

[130] S. R. Sharpe and Jr Singleton, R. L. Spontaneous flavor and parity breaking with Wilson fermions. *Phys.Rev.*, D58:074501, 1998.

[131] P. H. Damgaard, K. Splittorff, and J. J. M. Verbaarschot. Microscopic spectrum of the Wilson Dirac operator. *Phys. Rev. Lett.*, 105:162002, 2010.

[132] R. Frezzotti and G. C. Rossi. Chirally improving Wilson fermions. 1. O(a) improvement. *JHEP*, 0408:007, 2004.

[133] G. Munster, C. Schmidt, and E. E. Scholz. Chiral perturbation theory for twisted mass QCD. *Nucl. Phys. Proc. Suppl.*, 140:320–322, 2005.

[134] R. Narayanan and H. Neuberger. A construction of lattice chiral gauge theories. *Nucl. Phys.*, B443:305–385, 1995.

[135] D. B. Kaplan. A method for simulating chiral fermions on the lattice. *Phys. Lett.*, B288:342–347, 1992.

[136] V. Furman and Y. Shamir. Axial symmetries in lattice QCD with Kaplan fermions. *Nucl. Phys.*, B439:54–78, 1995.

[137] M. Creutz. Chiral anomalies and rooted staggered fermions. *Phys. Lett.*, B649:230–234, 2007.

[138] M. Creutz. Four-dimensional graphene and chiral fermions. *JHEP*, 04:017, 2008.

[139] T. Misumi, M. Creutz, and T. Kimura. Classification and generalization of minimal-doubling actions. *PoS*, LATTICE2010:260, 2010.

[140] L. H. Karsten. Lattice fermions in euclidean space-time. *Phys. Lett.*, B104:315, 1981.

[141] F. Wilczek. On lattice fermions. *Phys. Rev. Lett.*, 59:2397, 1987.

[142] A. Borici. Creutz fermions on an orthogonal lattice. *Phys. Rev.*, D78:074504, 2008.

[143] P. F. Bedaque, M. I. Buchoff, B. C. Tiburzi, and A. Walker-Loud. Search for fermion actions on hyperdiamond lattices. *Phys. Rev.*, D78:017502, 2008.

[144] T. Kimura and T. Misumi. Lattice fermions based on higher-dimensional hyperdiamond lattices. *Prog. Theor. Phys.*, 123:63–78, 2010.

[145] M. Creutz and T. Misumi. Classification of minimally doubled fermions. *Phys. Rev.*, D82:074502, 2010.

[146] J. H. Weber. *Properties of minimally doubled fermions*. PhD thesis, 2017.

[147] M. Creutz. Why rooting fails. *PoS*, LAT2007:007, 2007.

[148] M. Creutz. Anomalies and discrete chiral symmetries. *PoS*, QCD-TNT09:008, 2009.

[149] S. Capitani, M. Creutz, J. Weber, and H. Wittig. Renormalization of minimally doubled fermions. *JHEP*, 09:027, 2010.

[150] B. C. Tiburzi. Chiral lattice fermions, minimal doubling, and the axial anomaly. *Phys. Rev.*, D82:034511, 2010.

[151] R. Narayanan and H. Neuberger. Chiral determinant as an overlap of two vacua. *Nucl. Phys.*, B412:574–606, 1994.

[152] W. Shockley. On the surface states associated with a periodic potential. *Phys. Rev.*, 56:317–323, 1939.

[153] C. G. Callan, Jr. and J. A. Harvey. Anomalies and fermion zero modes on strings and domain walls. *Nucl. Phys.*, B250:427–436, 1985.

[154] K. Jansen. Domain wall fermions and chiral gauge theories. *Phys. Rept.*, 273:1–54, 1996.

[155] M. Creutz and I. Horvath. Surface states and chiral symmetry on the lattice. *Phys. Rev.*, D50:2297–2308, 1994.

[156] Y. Shamir. Chiral fermions from lattice boundaries. *Nucl. Phys.*, B406: 90–106, 1993.

[157] R. Narayanan and H. Neuberger. Infinitely many regulator fields for chiral fermions. *Phys. Lett.*, B302:62–69, 1993.

[158] H. Neuberger. A practical implementation of the overlap Dirac operator. *Phys. Rev. Lett.*, 81:4060–4062, 1998.

[159] H. Neuberger. More about exactly massless quarks on the lattice. *Phys. Lett.*, B427:353–355, 1998.

[160] T.-W. Chiu and S. V. Zenkin. On solutions of the Ginsparg-Wilson relation. *Phys. Rev.*, D59:074501, 1999.

[161] S. Chandrasekharan. Lattice QCD with Ginsparg-Wilson fermions. *Phys. Rev.*, D60:074503, 1999.

[162] P. Hasenfratz, V. Laliena, and F. Niedermayer. The index theorem in QCD with a finite cutoff. *Phys. Lett.*, B427:125–131, 1998.

[163] P. Hasenfratz. Lattice QCD without tuning, mixing and current renormalization. *Nucl. Phys.*, B525:401–409, 1998.

[164] M. Creutz. Transiting topological sectors with the overlap. *Nucl. Phys. Proc. Suppl.*, 119:837–839, 2003.

[165] M. Luscher. Exact chiral symmetry on the lattice and the Ginsparg-Wilson relation. *Phys. Lett.*, B428:342–345, 1998.

[166] H. Neuberger. Overlap. *Chin. J. Phys.*, 38:533–542, 2000.

[167] P. Hernandez, K. Jansen, and M. Luscher. Locality properties of Neuberger's lattice Dirac operator. *Nucl. Phys.*, B552:363–378, 1999.

[168] I. Horvath. Ginsparg-Wilson relation and ultralocality. *Phys. Rev. Lett.*, 81:4063–4066, 1998.

[169] I. Horvath. Ginsparg-Wilson-Luscher symmetry and ultralocality. *Phys. Rev.*, D60:034510, 1999.

[170] R. G. Edwards, U. M. Heller, and R. Narayanan. A study of chiral symmetry in quenched QCD using the overlap Dirac operator. *Phys. Rev.*, D59:094510, 1999.

[171] A. Borici. Lanczos approach to the inverse square root of a large and sparse matrix. *J. Comput. Phys.*, 162:123–131, 2000.

[172] S. J. Dong, F. X. Lee, K. F. Liu, and J. B. Zhang. Chiral symmetry, quark mass, and scaling of the overlap fermions. *Phys. Rev. Lett.*, 85:5051–5054, 2000.

[173] C. Gattringer. A new approach to Ginsparg-Wilson fermions. *Phys. Rev.*, D63:114501, 2001.

[174] M. Luscher. Lattice regularization of chiral gauge theories to all orders of perturbation theory. *JHEP*, 0006:028, 2000.

[175] Y. Kikukawa and A. Yamada. Weak coupling expansion of massless QCD with a Ginsparg-Wilson fermion and axial U(1) anomaly. *Phys. Lett.*, B448:265–274, 1999.

[176] G. 't Hooft. Naturalness, chiral symmetry, and spontaneous chiral symmetry breaking. *NATO Adv. Study Inst. Ser. B Phys.*, 59:135, 1980.

[177] R. G. Edwards, U. M. Heller, and R. Narayanan. Evidence for fractional topological charge in SU(2) pure Yang-Mills theory. *Phys. Lett.*, B438: 96–98, 1998.

[178] L. H. Karsten and J. Smit. Lattice fermions: species doubling, chiral invariance, and the triangle anomaly. *Nucl. Phys.*, B183:103, 1981.

[179] J. B. Kogut and L. Susskind. Hamiltonian formulation of Wilson's lattice gauge theories. *Phys. Rev.*, D11:395, 1975.

[180] L. Susskind. Lattice fermions. *Phys. Rev.*, D16:3031–3039, 1977.

[181] H. S. Sharatchandra, H. J. Thun, and P. Weisz. Susskind fermions on a Euclidean lattice. *Nucl. Phys.*, B192:205, 1981.

[182] Y. Aoki, Z. Fodor, S. D. Katz, and K. K. Szabo. The QCD transition temperature: Results with physical masses in the continuum limit. *Phys. Lett.*, B643:46–54, 2006.

[183] M. Cheng, N. H. Christ, S. Datta, J. van der Heide, C. Jung, *et al.* The QCD equation of state with almost physical quark masses. *Phys. Rev.*, D77:014511, 2008.

[184] A. Bazavov, D. Toussaint, C. Bernard, J. Laiho, C. DeTar, *et al.* Nonperturbative QCD simulations with 2+1 flavors of improved staggered quarks. *Rev. Mod. Phys.*, 82:1349–1417, 2010.

[185] E. Lunghi and A. Soni. Possible evidence for the breakdown of the CKM-paradigm of CP-violation. *Phys. Lett.* B697:323–328, 2011.

[186] H. Hamber and G. Parisi. Numerical estimates of hadronic masses in a pure SU(3) gauge theory. *Phys. Rev. Lett.*, 47:1792, 1981. [,619(1981)].

[187] M. Creutz. Anomalies, gauge field topology, and the lattice. *Annals Phys.*, 326:911–925, 2011.

[188] M. Teper. Instantons in the quantized su(2) vacuum: A lattice Monte Carlo investigation. *Phys. Lett.*, B162:357, 1985.

[189] M. Luscher. Topology of lattice gauge fields. *Commun. Math. Phys.*, 85:39, 1982.

[190] F. Bruckmann, C. Gattringer, E.-M. Ilgenfritz, M. Muller-Preussker, A. Schafer, *et al.* Quantitative comparison of filtering methods in lattice QCD. *Eur. Phys. J.*, A33:333–338, 2007.

[191] M. Luscher. Properties and uses of the Wilson flow in lattice QCD. *JHEP*, 1008:071, 2010.

[192] F. Bruckmann, F. Gruber, K. Jansen, M. Marinkovic, C. Urbach, *et al.* Comparing topological charge definitions using topology fixing actions. *Eur. Phys. J.*, A43:303–311, 2010.

[193] P. J. Moran, D. B. Leinweber, and J. Zhang. Wilson mass dependence of the overlap topological charge density. *Phys. Lett.*, B695:337–342, 2011.

[194] M. Luscher. Topological effects in QCD and the problem of short distance singularities. *Phys. Lett.*, B593:296–301, 2004.

[195] L. Giusti, G. C. Rossi, and M. Testa. Topological susceptibility in full QCD with Ginsparg-Wilson fermions. *Phys. Lett.*, B587:157–166, 2004.

[196] M. Luscher and F. Palombi. Universality of the topological susceptibility in the SU(3) gauge theory. *JHEP*, 1009:110, 2010.

[197] E. Witten. Current algebra theorems for the $u(1)$ Goldstone boson. *Nucl. Phys.*, B156:269, 1979.

[198] G. Veneziano. $u(1)$ without instantons. *Nucl. Phys.*, B159:213–224, 1979.

[199] E. Eichten and J. Preskill. Chiral gauge theories on the lattice. *Nucl. Phys.*, B268:179, 1986.

[200] E. Poppitz and Y. Shang. Chiral lattice gauge theories via mirror-fermion decoupling: a mission (im)possible? *Int. J. Mod. Phys.*, A25:2761–2813, 2010.

[201] I. Montvay. A chiral SU(2)-L x SU(2)-R gauge model on the lattice. *Phys. Lett.*, B199:89, 1987.

[202] M. Creutz. Is the doubler of the electron an antiquark? *Nucl. Phys. Proc. Suppl.*, 63:599–601, 1998.

[203] M. F. L. Golterman, K. Jansen, D. N. Petcher, and J. C. Vink. Investigation of the domain wall fermion approach to chiral gauge theories on the lattice. *Phys. Rev.*, D49:1606–1620, 1994.

[204] H. Neuberger. Comments on 'Lattice formulation of the standard model' by Creutz, *et al. Phys. Lett.*, B413:387–390, 1997.

[205] E. H. Fradkin and S. H. Shenker. Phase diagrams of lattice gauge theories with Higgs fields. *Phys. Rev.*, D19:3682–3697, 1979.

Index